# The Economic and Social Impacts of E-Commerce

Sam Lubbe
Cape Technikon, South Africa

Johanna Maria van Heerden
JS Consultants, South Africa

**IDEA GROUP PUBLISHING**

Hershey • London • Melbourne • Singapore • Beijing

381.1
L92e

| | |
|---|---|
| Acquisition Editor: | Mehdi Khosrow-Pour |
| Senior Managing Editor: | Jan Travers |
| Managing Editor: | Amanda Appicello |
| Development Editor: | Michele Rossi |
| Copy Editor: | Terry Heffelfinger |
| Typesetter: | Amanda Lutz |
| Cover Design: | Kory Gongloff |
| Printed at: | Integrated Book Technology |

Published in the United States of America by
    Idea Group Publishing (an imprint of Idea Group Inc.)
    701 E. Chocolate Avenue
    Hershey PA 17033
    Tel: 717-533-8845
    Fax: 717-533-8661
    E-mail: cust@idea-group.com
    Web site: http://www.idea-group.com

and in the United Kingdom by
    ᴊₑ Idea Group Publishing (an imprint of Idea Group Inc.)
    3 Henrietta Street
    Covent Garden
    London WC2E 8LU
    Tel: 44 20 7240 0856
    Fax: 44 20 7379 3313
    Web site: http://www.eurospan.co.uk

Copyright © 2003 by Idea Group Inc. All rights reserved. No part of this book may be reproduced in any form or by any means, electronic or mechanical, including photocopying, without written permission from the publisher.

Library of Congress Cataloging-in-Publication Data

Lubbe, Sam, 1952-
   The economic and social impacts of e-commerce / Sam Lubbe.
       p. cm.
Includes bibliographical references (p.  ) and index.
   ISBN 1-59140-043-0 (hard cover) -- ISBN 1-59140-077-5 (ebook)
   1. Electronic commerce. 2. Electronic commerce--South Africa. I.
Title.
   HF5548.32.L82 2003
   381'.1--dc21
                                                        2002156242


British Cataloguing in Publication Data
A Cataloguing in Publication record for this book is available from the British Library.

# *NEW* from Idea Group Publishing

- **Digital Bridges: Developing Countries in the Knowledge Economy**, John Senyo Afele/ ISBN:1-59140-039-2; eISBN 1-59140-067-8, © 2003
- **Integrative Document & Content Management: Strategies for Exploiting Enterprise Knowledge**, Len Asprey and Michael Middleton/ ISBN: 1-59140-055-4; eISBN 1-59140-068-6, © 2003
- **Critical Reflections on Information Systems: A Systemic Approach**, Jeimy Cano/ ISBN: 1-59140-040-6; eISBN 1-59140-069-4, © 2003
- **Web-Enabled Systems Integration: Practices and Challenges**, Ajantha Dahanayake and Waltraud Gerhardt ISBN: 1-59140-041-4; eISBN 1-59140-070-8, © 2003
- **Public Information Technology: Policy and Management Issues**, G. David Garson/ ISBN: 1-59140-060-0; eISBN 1-59140-071-6, © 2003
- **Knowledge and Information Technology Management: Human and Social Perspectives**, Angappa Gunasekaran, Omar Khalil and Syed Mahbubur Rahman/ ISBN: 1-59140-032-5; eISBN 1-59140-072-4, © 2003
- **Building Knowledge Economies: Opportunities and Challenges**, Liaquat Hossain and Virginia Gibson/ ISBN: 1-59140-059-7; eISBN 1-59140-073-2, © 2003
- **Knowledge and Business Process Management**, Vlatka Hlupic/ISBN: 1-59140-036-8; eISBN 1-59140-074-0, © 2003
- **IT-Based Management: Challenges and Solutions**, Luiz Antonio Joia/ISBN: 1-59140-033-3; eISBN 1-59140-075-9, © 2003
- **Geographic Information Systems and Health Applications**, Omar Khan/ ISBN: 1-59140-042-2; eISBN 1-59140-076-7, © 2003
- **The Economic and Social Impacts of E-Commerce**, Sam Lubbe/ ISBN: 1-59140-043-0; eISBN 1-59140-077-5, © 2003
- **Computational Intelligence in Control,** Masoud Mohammadian, Ruhul Amin Sarker and Xin Yao/ISBN: 1-59140-037-6; eISBN 1-59140-079-1, © 2003
- **Decision-Making Support Systems: Achievements and Challenges for the New Decade**, M.C. Manuel Mora, Guisseppi Forgionne and Jatinder N.D. Gupta/ISBN: 1-59140-045-7; eISBN 1-59140-080-5, © 2003
- **Architectural Issues of Web-Enabled Electronic Business**, Nansi Shi and V.K. Murthy/ ISBN: 1-59140-049-X; eISBN 1-59140-081-3, © 2003
- **Adaptive Evolutionary Information Systems**, Nandish V. Patel/ISBN: 1-59140-034-1; eISBN 1-59140-082-1, © 2003
- **Managing Data Mining Technologies in Organizations: Techniques and Applications**, Parag Pendharkar/ ISBN: 1-59140-057-0; eISBN 1-59140-083-X, © 2003
- **Intelligent Agent Software Engineering**, Valentina Plekhanova/ ISBN: 1-59140-046-5; eISBN 1-59140-084-8, © 2003
- **Advances in Software Maintenance Management: Technologies and Solutions**, Macario Polo, Mario Piattini and Francisco Ruiz/ ISBN: 1-59140-047-3; eISBN 1-59140-085-6, © 2003
- **Multidimensional Databases: Problems and Solutions**, Maurizio Rafanelli/ISBN: 1-59140-053-8; eISBN 1-59140-086-4, © 2003
- **Information Technology Enabled Global Customer Service**, Tapio Reponen/ISBN: 1-59140-048-1; eISBN 1-59140-087-2, © 2003
- **Creating Business Value with Information Technology: Challenges and Solutions**, Namchul Shin/ISBN: 1-59140-038-4; eISBN 1-59140-088-0, © 2003
- **Advances in Mobile Commerce Technologies**, Ee-Peng Lim and Keng Siau/ ISBN: 1-59140-052-X; eISBN 1-59140-089-9, © 2003
- **Mobile Commerce: Technology, Theory and Applications**, Brian Mennecke and Troy Strader/ ISBN: 1-59140-044-9; eISBN 1-59140-090-2, © 2003
- **Managing Multimedia-Enabled Technologies in Organizations**, S.R. Subramanya/ISBN: 1-59140-054-6; eISBN 1-59140-091-0, © 2003
- **Web-Powered Databases**, David Taniar and Johanna Wenny Rahayu/ISBN: 1-59140-035-X; eISBN 1-59140-092-9, © 2003
- **E-Commerce and Cultural Values**, Theerasak Thanasankit/ISBN: 1-59140-056-2; eISBN 1-59140-093-7, © 2003
- **Information Modeling for Internet Applications**, Patrick van Bommel/ISBN: 1-59140-050-3; eISBN 1-59140-094-5, © 2003
- **Data Mining: Opportunities and Challenges**, John Wang/ISBN: 1-59140-051-1; eISBN 1-59140-095-3, © 2003
- **Annals of Cases on Information Technology** – vol 5, Mehdi Khosrowpour/ ISBN: 1-59140-061-9; eISBN 1-59140-096-1, © 2003
- **Advanced Topics in Database Research** – vol 2, Keng Siau/ISBN: 1-59140-063-5; eISBN 1-59140-098-8, © 2003
- **Advanced Topics in End User Computing** – vol 2, Mo Adam Mahmood/ISBN: 1-59140-065-1; eISBN 1-59140-100-3, © 2003
- **Advanced Topics in Global Information Management** – vol 2, Felix Tan/ ISBN: 1-59140-064-3; eISBN 1-59140-101-1, © 2003
- **Advanced Topics in Information Resources Management** – vol 2, Mehdi Khosrowpour/ ISBN: 1-59140-062-7; eISBN 1-59140-099-6, © 2003

*Excellent additions to your institution's library! Recommend these titles to your Librarian!*

To receive a copy of the Idea Group Publishing catalog, please contact (toll free) 1/800-345-4332,
fax 1/717-533-8661,or visit the IGP Online Bookstore at:
[http://www.idea-group.com]!
Note: All IGP books are also available as ebooks on netlibrary.com as well as other ebook sources.
Contact Ms. Carrie Skovrinskie at [cskovrinskie@idea-group.com] to receive a complete list of sources
where you can obtain ebook information or IGP titles.

University Libraries
Carnegie Mellon University
Pittsburgh, PA 15213-3890

# The Economic and Social Impacts of e-Commerce

## Table of Contents

Preface ........................................................................................ vi

Chapter I. TrickE-Business: Malcontents in the Matrix .................... 1
*Paul Taylor, University of Leeds, UK*

Chapter II. The Economic and Social Impact of Electronic Commerce in
Developing Countries ......................................................... 22
*Roberto Vinaja, University of Texas, Pan America, USA*

Chapter III. Adverse Effects of E-Commerce ................................... 33
*Sushil K. Sharma, Ball State University, USA*
*Jatinder N. D. Gupta, University of Alabama in Huntsville, USA*

Chapter IV. The Emerging Need for E-Commerce Accepted Practice
(ECAP) ............................................................................. 50
*G. Erwin, Cape Technikon, South Africa*
*S. Singh, University of South Africa, South Africa*

Chapter V. The Theory Behind the Economic Role of Managing the
Strategic Alignment of Organizations while Creating
New Markets ..................................................................... 69
*Sam Lubbe, Cape Technikon, South Africa*

Chapter VI. Online Customer Service ............................................. 95
*Rick Gibson, American University, USA*

Chapter VII. E-Commerce and Executive Information Systems:
A Managerial Perspective ................................................. 103
*G. Erwin, Cape Technikon, South Africa*
*Udo Averweg, University of Natal, South Africa*

**Chapter VIII. SMEs in South Africa: Acceptance and Adoption of E-Commerce** ......................................................... 121

*Eric Cloete, University of Cape Town, South Africa*

**Chapter IX. Key Indicators for Successful Internet Commerce: A South African Study** ................................................ 135

*Sam Lubbe, Cape Technikon, South Africa*
*Shaun Pather, Cape Technikon, South Africa*

**Chapter X. E-Learning is a Social Tool for E-Commerce at Tertiary Institutions** ........................................................ 154

*Marlon Parker, Cape Technikon, South Africa*

**Chapter XI. Relating Cognitive Problem-Solving Style to User Resistance** .............................................................. 184

*Michael Mullany, Northland Polytechnic, New Zealand*
*Peter Lay, Northland Polytechnic, New Zealand*

**Chapter XII. Electronic Commerce and Data Privacy: The Impact of Privacy Concerns on Electronic Commerce Use and Regulatory Preferences** ................................................................. 213

*Sandra C. Henderson, Auburn University, USA*
*Charles A. Snyder, Auburn University, USA*
*Terry Anthony Byrd, Auburn University, USA*

**Chapter XIII. Impersonal Trust in B2B Electronic Commerce: A Process View** ....................................................... 239

*Paul A. Pavlou, University of Southern California, USA*

**About the Authors** ...................................................... 258

**Index** ............................................................................ 263

# Preface

E-commerce is not new, though the interest shown in it is of relatively recent origin. Academics have applied their skill in seeking to maintain or improve business efficiency for years past, but they have concerned themselves mainly with obtaining facts of a historical nature – that is, by analyzing past papers, they have sought to regulate future policies. Until more recently they have been chiefly occupied with matters of a domestic or internal nature, and although they have not been able to ignore affairs outside, such as the influence exerted by customers, nevertheless they have not sought to extend the field of their activities. They have concentrated their endeavors on seeking to establish an efficiently run business, leaving those engaged on the various executive activities of the organization to pronounce on their own particular fields of interest.

Modern business activities and the increasing complexity of present-day e-commerce have necessitated a broadening of the views, knowledge and influence of the consultant, and while greater specialization has taken place within the profession itself, a new branch of IT has evolved, namely, that of e-commerce.

"E-Commerce" may be defined broadly as that aspect of IT that is concerned with the efficient management of a business through the presentation to management of such information as will facilitate efficient and opportune planning and control.

The managerial aspect of his work is the management consultant's prime concern. Having satisfied himself as to the efficiency of the organization of the business – covering such matters as the regulation of activities – he may justifiably expect to be concerned with the day-to-day running of affairs. His attention should be directed more particularly towards the extraction of information from records and the compilation and preparation of statements that will enable management to function with the minimum of effort and with the maximum of efficiency.

The term *e-commerce* has been used carefully for the title of this book, because it covers a broader view than "e-commerce." To carry out his duties effectively, the manager is now required to extend his knowledge and research into related but distinct fields of activity covering disparate areas such as taxation, manufacturing processes, electronic data processing, stock exchange activities, economic influences and statistical research.

The consultant needs not to be (for example) a qualified production engineer to be able to concern himself with the efficiency of e-commerce processes, but he must nevertheless have some knowledge of the organization's workflow to be able to assess what costing, statistical or other records are necessary to ensure effective control. His training and experience must enable him to comprehend and deal with these allied activities.

The essential characteristics of information required for e-commerce management are that (a) it must be relevant, and (b) it must be timely. To meet the first requirement, the e-consultant needs to have a detailed understanding of the business concerned. They must also have the ability to present such information in a way that enables management to concentrate on essential matters. The ideal e-consultant presents information to management without wasting time on routine activities that were previously assessed and concurred. It is here that "management by exception" should be operated. At the same time, if capital projects, expansion or proposed mergers are under consideration, it will be the manager's duty to grasp the underlying essentials of the situation and to present them in a way that will enable management to reach a decision based on all relevant facts.

In the second instance, the e-commerce manager must realize that information, to be useful, should be received in enough time to enable the executive to act effectively. To be informed after events have reached a stage that precludes their regulation or adjustment merely causes frustration and may lead to wrong decisions, aggravating an already difficult situation. It is here that factors of planning and control manifest themselves as essential to sound management. In analyzing the functions of the accountant regarding his presentation of information to management, his duties may be sub-divided as to:

1.  The presentation of forecasts and budgets of a forward-looking nature, facilitating planning.
2.  The supplying of such current information will ensure efficient control of activities during the fulfillment of the plans formulated.
3.  Ensuring that internal control within the business is such that relevant information is automatically prepared and summarized in such a way as provides an easy, rapid analysis and compilation for submission to management.

The application of control, particularly *flexible* control, presupposes the availability of sufficient information being at hand for budgeting. An efficient office routine is essential, as mentioned previously, but – and here the wider aspects of the e-commerce's experience must be applied – relevant information must also be made available from the web floor, the warehouse and the sales department. Part of such information should arise in the routine order of work, for example, the preparation of requisitions and their subsequent analysis or of efficient stock recording. Other information would have to be prepared specially, as for example, sales budgets and market analysis. In any case, the accountant should know what kind of information is likely to be useful and should ensure it will be received in time for analysis, interpretation and presentation to management.

Nevertheless, the broader aspects of planning will no doubt require the preparation of statistics and the amassing of information in those wider aspects of the e-commerce manager's field of experience. Where projects are to be undertaken, not only will a recommendation as to an adequate return on capital invested be required, but also the most suitable method of raising the necessary finance will have to be indicated. Likewise, if any take-over project or investment in a subsidiary company or other concern is contemplated, e-commerce will be expected to be able to express an opinion based upon the ability to interpret accounts, to assess future trading prospects, etc.

The importance of information being received in time for effective action has already been stressed. In this respect, the submission of information covering standards and variances from those standards during the course of actual activities will facilitate management by exception and effective action while control may still be exercised.

The necessity for the efficient recording of essential information has already been dealt with. This assumes efficient internal control and the suitable allocation of duties within the e-commerce's department so information may be rapidly compiled in an orderly manner, especially in the event of some urgent business arising, ensuring that no dislocation occurs.

Being in the nature of an introduction to the field of e-commerce, this brief exposition has sought only to illuminate some of the main aspects of the subject and to emphasize the duties falling to the management consultant; the more detailed aspects are dealt with in the pages of this book.

In the first chapter, Dr. Paul Taylor describes from a 'Devil's Advocate' stance the cultural context to the rise of various online activities that oppose the general values of e-Business. In the new digital times, capitalism's iconoclastic qualities have been enthusiastically re-appropriated by business gurus on the opposite side of the political spectrum.

In the second chapter, Roberto Vinaja addresses the potential benefits of Electronic Commerce to developing countries. Electronic commerce has many potential benefits for developing countries (DC). In his chapter, he describes the potential benefits of Electronic Commerce for developing nations and he provides case examples that illustrate this trend. The widespread adoption of electronic commerce is especially important for developing countries.

Jatinder Gupta describes in the third chapter the various adverse effects that have accompanied the advent of the Internet and e-commerce revolution. The Internet has become an incredibly powerful tool for conducting business electronically. Companies have taken the proactive approach and are jumping on the new way to conduct business. E-commerce greatly enables organizational change and helps organizations to conduct business with improved efficiencies and productivity. E-commerce is credited with empowering employees and knowledge workers in particular, by giving them easy access to virtually unlimited information. E-commerce technologies have helped nations to accelerate their economic growth and to provide more opportunities for the businesses to grow. Meanwhile, it has also created many challenges and adverse effects, such as concerns over privacy, consumer protection, security of credit card purchases, displacement of workers (especially low-status ones) and a negative quality of work life.

In the fourth chapter, Geoff Erwin shows that with the proliferation of the Internet and constant technological advancements, e-commerce will reshape the business world. Government organizations, large co-operations, medium and small business will have to organize their information and information systems in an accountable, well-structured way. He also asks "How do we document electronic businesses activities?"

In the fifth chapter, Sam Lubbe notes that the economic impact on e-commerce is and how this could be used to create new markets and to improve the strategic alignment of the organization. Over the past couple of years, the Internet has taken off and organizations will soon reap economic benefits on it. E-commerce will therefore hopefully emerge as an efficient yet effective mode of creating new markets, although most managers still doubt the economic impact and profitability it has. Enabled by global telecommunication networks and the convergence of computing, telecom, entertainment and publishing industries, e-commerce is supplanting (maybe replacing) traditional commerce. In the process, it is creating new economic opportunities for today's businesses, creating new market structures. Managers of tomorrow must therefore understand what e-commerce is; how the approach to this concept will be; and how it will affect the economic position of the organization. These questions could therefore be asked: What is the return on investment (ROI) on e-commerce? What is the effect of e-commerce on the strategic alignment of the organization? What is the economic effect of the strategic alignment on the organization?

In the sixth chapter, Rick Gibson looks into an effective online customer service strategy. Although the effectiveness of the online customer service will vary and depend on the type of business the company is involved in, the usage of different types of tools in this arena have proven to be more useful than others. Effectiveness in this work will be used in the sense that the more effective strategy will lead to more satisfied customers, a higher customer retention rate and higher revenue for the business.

In the seventh chapter, Geoff Erwin relates to the fact that Executive Information Systems (EIS) are designed to serve the needs of executive users in strategic planning and decision-making and for making both strategic and tactical decisions. The accessibility, navigation and management of data and information for improved executive decision-making are becoming *critical* in the new global business environment.

In the eighth chapter, Eric Cloete addresses how these small businesses in a developing country perceive the potential benefits of e-commerce and look at their consequent adoption of e-commerce activities in their own organizations. Comparisons are made between studies conducted in first world countries, particularly regarding the role of government initiatives.

In the ninth chapter, Shaun Pather and Sam Lubbe address the fact that the world of Internet commerce has been rapidly evolving since its advent in the 1990s.. This has had implications on research directions in the field of Electronic Commerce (e-commerce). No longer is it sufficient to study the formation of electronic markets in e-commerce. It is also necessary to have insight into the electronic markets' innermost workings.

In chapter ten, Marlon Parker states that tertiary education institutions aim to be recognized for social, knowledge and economic contributions in South Africa. There has also been an increase in the different uses (including e-learning) of the Internet. This increase has contributed to the electronic learning revolution and some South African tertiary institutions are making the technology-based paradigm shift for this reason.

In the eleventh chapter, Michael Mullany and Peter Lay investigated the relationships between user resistance to new information systems (such as e-commerce) and the differences in cognitive problem-solving styles between systems developers (analysts) and users.

In chapter twelve, Sandra Henderson, Charles Snyder and Terry Byrd present a study examining the relationships between consumer privacy concerns, actual e-commerce activity, the importance of privacy policies and regulatory preference.

In the final chapter, Paul Pavlou addresses the issue of "impersonal trust" in establishing successful B2B relationships–the type of trust that is created by structural arrangements, rather than from repeated interaction and familiarity.

# Acknowledgments

E-commerce has made significant progress even in the short space of time before this book was published, and the importance of information for management purposes has become more widely appreciated. In this edition, therefore, we have incorporated illustrations of the types of research areas likely to facilitate the formulation of economic and social management policies. The section on social statements and economic impact has been given extensive treatment and matters covering e-commerce valuation and the understanding of the aspects of finance have been brought up to date.

A special project of a practical nature has been introduced to demonstrate the compilation and application of economic principles to emphasize the essential role to co-ordinate all the e-functions of the business. Greater recognition of the usefulness of sources, properly applied, warrants fuller treatment of this subject, while the opportunity has been taken to include the latest recommendations of e-commerce researchers.

This book follows the recommendations of the various e-Initiative bodies and of the bodies responsible for further research.

Acknowledgments are due to those who have written offering chapters, their appreciation and suggestions for improvement.

*Sam Lubbe*
*June 2002*

Chapter I

# TrickE-Business: Malcontents in the Matrix

Paul A. Taylor
University of Leeds, UK

## ABSTRACT

*This chapter explores the phenomenon of hacktivism in the context of globalization debates and the evolving nature of new social movements. It explores the historical trend by which capitalism has become increasingly more immaterial in its appearance but powerful in its effects. Using examples of specific hacktivist groups, hacktivism is shown to be an inventive response to this trend and represents an imaginative re-appropriation of the Web for spider-like anti-capitalist protest. The paper concludes with a summary of the hacktivist philosophy that seeks to reassert the origins of the marketplace as an agora for the people rather than just big business. Hacktivism is shown to represent a rationale diametrically opposed to e-commerce.*

## INTRODUCTION-
## ALL THAT IS SOLID MELTS INTO AIR...

*Constant revolutionizing of production, uninterrupted disturbance of all social relations, everlasting uncertainty and agitation, distinguish the bourgeois epoch from all earlier times. All fixed, fast-frozen relationships, with their train of venerable ideas and opinions, are swept away, all new-formed ones become obsolete before they can ossify. All that is solid melts into air, all that is holy is*

Copyright © 2003, Idea Group Inc. Copying or distributing in print or electronic forms without written permission of Idea Group Inc. is prohibited.

*profaned.* (Marx & Engels – The Manifesto of the Communist Party[1])

*The Robespierre of this revolution is finance capital ... As the Jacobins learned during the French Revolution, it is the most zealous, principled advocates of new values who are ultimately most at risk in a revolutionary environment.* (Greider, 1997: 25, 26)

The purpose of this chapter is to describe from a 'Devil's Advocate' stance the cultural context to the rise of various online activities that oppose the general values of e-business. In the new digital times, Marx's description of capitalism's iconoclastic qualities has been enthusiastically re-appropriated by business gurus on the opposite side of the political spectrum. His criticism of disorienting change has been swamped by a tsunami of techno-enthusiasm. The perennial pertinence of Marx's poetically-charged analysis of the socially transformative power of capitalism's increasingly immaterial form is illustrated in a spate of such recently evocative titles as: *Living on Thin Air,' The Empty Raincoat,' Being Digital;* and *The Weightless World.* Such New Economy tracts can even make Marx's florid language seem relatively understated - to the extent that it has been described as the 'deranged optimism' and 'corporate salivating' of 'business pornography' (Thomas Frank 2001). In this atmosphere of revolutionary rhetoric, however, Greider's above quotation hints at the dangers that can await those at the vanguard of change. We will see later in this chapter that just as Marx argued that capitalism contained its own fatal internal contradictions, so various writers are beginning to argue that the technological infrastructure of e-commerce may provide the fertile grounds for oppositional forces.

The dot.com revolution has produced dot.communists, and in addition to the recent slowdown in the revolution's own internal momentum, the information superhighway now has speed bumps in the form of online political activists known as hacktivists. Together, hacktivists and anti-corporate theorists are creating a groundswell of opinion that may mitigate future growth in e-commerce and the dream of abstract friction-free capitalism.

# THE MANIFEST DESTINY OF FRICTION-FREE CAPITALISM

*Now capital has wings* – (New York financier Robert A. Johnson)[2].

Copyright © 2003, Idea Group Inc. Copying or distributing in print or electronic forms without written permission of Idea Group Inc. is prohibited.

*For how many eons had insurmountable geography impeded man's business? Now the new American race had burst those shackles. Now it could couple its energies in one overarching corporation, one integrated instrument of production whose bounty might grow beyond thwarting.* (Powers 1998: 91)

According to Brown (1998), The phrase *manifest destiny* was coined by John L. O'Sullivan, the editor of the United States Magazine and Demographic Review (July-August 1845), when he said that opposition to the U.S. takeover of Texas from Mexico interfered with "the fulfilment of our manifest destiny to overspread the continent allotted by Providence for the free development of our yearly multiplying millions" (Brown 1998: 2). It has subsequently been used for many years to encapsulate the expansive mentality of U.S. foreign policy. In a post-Cold War international environment where U.S. economic dominance has increasingly supplanted overt military force as its primary source of global influence, manifest destiny is a freshly evocative concept that encapsulates the expansionary and evangelical nature of a global economic order driven by American values:

> One memorable incident, at a meeting of economic policy-makers from the largest industrialized countries that was held in Denver in June 1997, signalled the new mood. President Clinton and Larry Summers, then deputy secretary of the treasury, seized the occasion to tell the world about the miraculous new American way. They handed out pairs of cowboy boots and proceeded to entertain the foreigners with what the Financial Times called a steady diet of "effusive self-praise" spiced with occasional "harsh words … for the rigidities of French and European markets." Don your boots and down with France! (Frank 2001:7)

The above account neatly conflates how the Wild West acts as trope for U.S. attitudes regarding globalization and the accompanying distaste that a gung-ho frontier attitude implies for those with less expansive attitudes more protective of cultural factors. Implied in this outlook is a world economic order viewed as virgin territory to be pioneered with a minimum of regulatory brakes. The key significance of the Wild West motif is the way that the decontextualized abstract space of the frontier replaces the messy contingencies of specific locales. The 'friction-free' capitalism that globalization is predicated upon replaces local concerns with more general, immaterial imperatives in a manner remarkably unchanged since it was described so forcefully in the *Manifesto of the Communist Party*:

> … the world-market [has] given a cosmopolitan character to produc-

Copyright © 2003, Idea Group Inc. Copying or distributing in print or electronic forms without written permission of Idea Group Inc. is prohibited.

tion and consumption in every country … it has drawn from under the feet of industry the national ground on which it stood. Industries … no longer work up indigenous raw material, but raw material drawn from the remotest zones; industries whose products are consumed, not only at home, but in every quarter of the globe … And as in material production, so also in intellectual production. The intellectual creations of individual nations become common property. National one-sidedness and narrow-mindedness become more and more impossible, and from the numerous national and local literatures, there arises a world literature. (Marx & Engels in Tucker 1978: 476-477)

The smooth, almost virus-like expansionary nature of globalized, de-localized capitalism is perhaps best illustrated by the notion of the franchise. The homogenous urban geography across the globe is testament to the ease with which commodities transcend cultural contexts, taking the golden arches of McDonalds to Moscow in a "three-ring binder" process as satirized in the cyberpunk novel *Snowcrash*:

The franchise and the virus work in the same principle; what thrives in one place will thrive in another. You just have to find a sufficiently virulent business plan, condense it into a three-ring binder – its DNA – Xerox it, and embed it in the fertile lining of a well-travelled highway, preferably one with a left-turn lane. Then the growth will expand until it runs up against its property lines (Stephenson 1992: 178).

Concern at the virulence with which the commodity form spreads into other cultures stems from its inherently abstract, context-free logic. There is a deeply embedded, cultural alignment between *laissez-faire* ideology and its heavily technologically mediated consumer products such as computing, Hollywood films, and fast-food franchises. The emblematic role of the latter has led to the adoption of the phrase 'the McDonaldization of …' to describe the application of corporate values to areas of life, such as the education sector, previously based upon a public service rather than commodity ethos. Freed from a grounded basis in a particular cultural context, the spread of corporate values assumes its own amoral expansionary raison d'être and a brutal end in itself, to the extent that Ray A. Kroc, the founder of McDonald's once said of his business rivals, "If they were drowning to death, I would put a hose in their mouth." (Schlosser 2001: 41) While this may be seen as an extreme, unrepresentative example of the corporate ethos, there is strong evidence to suggest that, at the very least, new technologies and expansionary business values have a tendency to align themselves to create a high degree of insensitivity to local context. Thus the McDonald's corporation has become one of

Copyright © 2003, Idea Group Inc. Copying or distributing in print or electronic forms without written permission of Idea Group Inc. is prohibited.

the world's leading purchasers of satellite imagery, using a software program called Quintillion to automate its site-selection process and the curator of the Holocaust museum at Dachau in southern Germany complained that the company distributed leaflets in the camp's car park: ' "Welcome to Dachau," said the leaflets, "and welcome to McDonalds." '(Schlosser 2001: 233).

The conjunction of a product's essentially homogenous nature, allied with such aggressively expansionist marketing techniques, and a disregard for local sensitivities is perhaps best captured by the 'clustering' strategy employed by Starbucks. Naomi Klein describes it in the following terms:

> Starbucks' policy is to drop "clusters" of outlets already dotted with cafes and espresso bars …Instead of opening a few stores in every city in the world, or even in North America, Starbucks waits until it can blitz an entire area and spread, to quote *Globe and Mail* columnist John Barber, like head lice through a kindergarten". (Klein 2000: 136)

This branding strategy is underpinned by a commitment to homogeneity that is succinctly captured in Theodore Levitt's (1983) essay, *The Globalization of Markets*, in which he advocated that: 'The global corporation operates with resolute constancy – at low relative cost – as if the entire world (or major regions of it) were a single entity; it sells the same things in the same way everywhere …Ancient differences in national tastes or modes of doing business disappear' (Levitt, 1983, cited in Klein 2000: 116). Moreover, homogenization extends beyond the heavily branded products of the global corporations. As franchises such as McDonald's and Starbucks spread throughout the world's cities, eliminating independent stores and smaller chains, there is an increased sense of 'sameness' about not only the content of the product, but also the urban environment within which it is provided. In other words, friction-free capitalism, encourages not only the standardization of product, but also the standardization of its surrounding environment, through the formation of what Deleuze (1989) refers to as *espace quelconque* or 'any-space-whatever'.

The departicularized, abstract spaces and flows upon which new information technologies and the e-boom are premised are particularly well-suited to this homogenizing quality of contemporary capitalism. Computer code utilizes abstract, digital representations of information to create generic models of reality to the extent that the words of Ellen Ullman, a U.S. computer programmer closely echo Deleuze's: "I begin to wonder if there isn't something in computer systems that is like a surburban development. Both take places - real, particular places - and turn them into anyplace." (Ullman 1997: 80) This generic, anyplace quality of computer code's binary digits is a specific technological manifestation of a more pervasively

Copyright © 2003, Idea Group Inc. Copying or distributing in print or electronic forms without written permission of Idea Group Inc. is prohibited.

experienced and commercially-induced aesthetic within society at large. Ullman, complains of the lack of rootedness and materiality of contemporary businesses to the extent that she thinks of: "The postmodern company as PC - a shell, a plastic cabinet. Let the people come and go; plug them in, then pull them out. (Ullman, 1997: 129) The rise in the profile of e-business, has taken place in this wider cultural climate of a generalized desire to abandon the particularities of the local and community ties for the abstractions Ullman describes. Klein (2001) refers to this process as *a race towards weightlessness* and it is the social consequences of such a race that we now address.

## E-COMMERCE AS EMPIRE & NEW SOCIAL MOVEMENTS

*Along with the global market and global circuits of production has emerged a global order, a new logic and structure of rule – in short, a new form of sovereignty. Empire is the political subject that effectively regulates these global exchanges, the sovereign power that governs the world.* (Hardt & Negri 2000: xi)

From a critical perspective, the transnational imperatives of global capitalism have spilled over from the world of business into the social realm. This has occurred in wide range of contexts. In the U.K., for example, Manchester United, the World's biggest football team has achieved that status by replacing its previous working class fan-base to become a global brand. Disengagement from historical social ties has culminated in the large recent "commercial tie-up" deal with the New York Yankees baseball team[3]. Meanwhile, in the field of politics, a similar loss of community-based activity is reflected in the Labour Party's *Operation Turnout*[4] for the UK's national election of 2001. This initiative takes the marketing ethos that created the soap-powder-sounding *New Labour* to its own logical branding conclusion by offering constituents a thirty-second doorstep chat with their MP, thereby inadvertently creating a pre-election version of the *Daz Doorstep Challenge*[5]. More generally, corporate values are now insinuated in areas of society previously protectively ring-fenced (even within neo-classical economics) by the concept of the 'public good'. Schools, universities, and hospitals, all now face centrally-imposed matrices of business-plans and statistical interrogations of performance.

In the eyes of capitalism's critics, new information technologies threaten to further engulf culture with corporate values: 'In the postmodernization of the global

Copyright © 2003, Idea Group Inc. Copying or distributing in print or electronic forms without written permission of Idea Group Inc. is prohibited.

economy, the creation of wealth tends ever more toward what we call biopolitical production, the production of social life itself, in which the economic, the political, and the cultural increasingly, overlap and invest one another.' (Hardt & Negri, 2000: xiii) The perception is that a corporate social environment has merged with a facilitative technical infrastructure to produce a culturally and technologically aligned informational matrix with abstract imperatives but very real effects. From league-tables to modularized, 'customer-orientated' university courses, the contemporary pervasiveness of corporate values is inextricably linked to new information technologies in a Microsoft Office-mentality that privileges the computer-mediated logic of efficiently specified means over normative discussions about desirable ends.

In the face of such global biopolitical forces, Hardt and Negri describe how a new form of social activism has arisen from a "paradox of incommunicability" (ibid: 54) and is characterized by two main properties:

1)   Each struggle starts at the local level, but jumps vertically to global attention.
2)   Struggles can increasingly be defined as "bio-political" because they blur the distinctions previously made between economics and politics and add the cultural to the new mix.

The paradox stems from the fact that despite living in a much heralded communication age, the local particularities of political struggles have become increasingly difficult to communicate between groups as the basis for any international chain of political action. Instead, such horizontal communication risks being supplanted by the increasing advent of "vertical events" such as the Tiananmen Square protests that jump into the global consciousness through the world's media.

Notwithstanding, Hardt and Negri's identification of the "vertical jump", increasing theoretical attention is being given to the ways in which the breaking of traditional "chains" of political protest have has created new horizontal modes of communication that seek to re-appropriate the ease with which global capital circulates its commodities and their values. Thus, Lash argues that: "With the dominance of communication there is a politics of struggle around not accumulation but *circulation*. Manufacturing capitalism privileges production and accumulation, the network society privileges communication and circulation." (Lash 2002: 112) Dyer-Witheford sees new contested sites of circulation: "the cyberspatial realm … increasingly provides a medium both for capitalist control and for the "circulation of struggles". (Dyer-Witheford, 1999: 13) Interesting questions are thus raised by the advent of new social movements that utilize a sophisticated *a priori* sense of circulation's importance. Within Weberian analysis the terms *Gesellschaft* and

Copyright © 2003, Idea Group Inc. Copying or distributing in print or electronic forms without written permission of Idea Group Inc. is prohibited.

*Gemeinschaft* are used to distinguish between feelings of belonging to an abstract society and a more intimate community respectively. New Web-based social movements have arguably produced a hybrid combination of both affinities. Their sense of belonging is abstract in the sense that it often refers to a sense of solidarity stretched by global distances, yet group solidarity is also nurtured by those same global communications that serve to reinforce awareness of the particularities of local struggles.

Social movements have become exactly that – movements - but often of socially relevant information rather than actual physical bodies of people (although the two categories may be combined in Web-facilitated protest events such as World Trade Organization demonstrations). Such new groups may be usefully understood as the affective groups Maffesoli (1996) describes as *neo-tribes*. In contrast to capitalism's "iron cage of rationality," new affective relationships are built upon a non-logical emotional basis, and in a more proactive version of Baudrillard's inertly fatal masses of postmodernity. For Maffesoli, such neo-tribes have a certain "underground puissance": 'The rational era is built on the principle of individuation and of separation, whereas the empathetic period is marked by the lack of differentiation, the "loss" in a collective subject: in other words, what I shall call neo-tribalism.' (Maffesoli 1996: 11) The new neo-tribes do not fit easily into the classificatory categories of the system that would absorb them: 'Their outlines are ill-defined: sex, appearance, lifestyles – even ideology – are increasingly qualified in terms ("trans", "meta") that go beyond the logic of identity and/or binary logic.' (ibid: 11) These new social movements ironically use the binary-based circulation systems of capitalism for their own non-binary purposes that in another semantic irony can perhaps be understood best in terms of a web.

# FROM NETWORKS TO WEBS

*The terminals of the network society are static. The bonding, on the other hand, of web weavers with machines is nomadic. They form communities with machines, navigate in cultural worlds attached to machines. These spiders weave not networks, but webs, perhaps electronic webs, undermining and undercutting the networks. Networks need walls. Webs go around the walls, up the walls, hide in the nooks and crannies and corners of where the walls meet ... Networks are shiny, new, flawless. Spiders' webs in contrast, attach to abandoned rooms, to disused objects, to the ruins, the disused and discarded objects of capitalist production. Networks are cast more or less in stone; webs are weak, easily destroyed.*

Copyright © 2003, Idea Group Inc. Copying or distributing in print or electronic forms without written permission of Idea Group Inc. is prohibited.

*Networks connect by a utilitarian logic, a logic of instrumental rationality. Webs are tactile, experiential rather than calculating, their reach more ontological than utilitarian.* (Lash 2002: 127)

In his *Practices of Everyday Life*, Michel de Certeau (1988) criticizes the expansionary nature of various systems of production that produce a society dominated by commodity value. He argues that resistance to such disciplining forces can be found in the various day-to-day subversions people carry out as they consume the products of such a dominant order. He uses the example of the indigenous Indians of South America who, although they superficially accepted the framework of the Catholic Church imposed upon them by the Spanish colonizers, in fact managed to develop various practices that kept their traditional values alive beneath the veneer of such acceptance and assimilation. In a similar fashion, he advocates the development of various strategies to resist the uniform, disciplinary effects of capitalism upon social life including the reappropriation of otherwise ordered urban environments in preference for more dynamic, liberated expressions of local particularities and interactions. De Certeau thus seeks escape routes from the circumscribing effects of the sorts of productive and organizational matrices previously described:

> We witness the advent of number. It comes with democracy, the large city, administrations, and cybernetics. It is a flexible and continuous mass, woven tight like a fabric with neither rips nor darned patches, a multitude of quantified heroes who lose names and faces as they become the ciphered river of the streets, a mobile language of computations and rationalities that belong to no one. (De Certeau, 1988: v)

De Certeau's identification of the tightly woven nature of fabric that accompanies "the advent of number" provides an earlier analysis of the subsequent focus upon capitalist networks such as that provided in the above quotation from Lash. Where De Certeau describes a cybernetic 'fabric with neither rips nor darned patches', Lash similarly talks of the 'flawless' nature of a utilitarian network. Lash proceeds to contrast the inherently disciplinary nature of such networks with the more organically libratory potential image of webs. He adopts Lefebvre's (1991) association of spiders' web making with the creation of autonomous spaces to make parallels with the potentially empowering web-forming activities of the new informational order's technoculture workers. In a very similar vein, Klein conceptualizes anti-corporate opposition as web-using spiders:

> … the image strikes me as a fitting one for this Web-age global activism. Logos, by the force of ubiquity, have become the closest thing we have

Copyright © 2003, Idea Group Inc. Copying or distributing in print or electronic forms without written permission of Idea Group Inc. is prohibited.

to an international language, recognized and understood in many more places than English. Activists are now free to swing off this web of logos like spy/spiders – trading information about labor practices, chemical spills, animal cruelty and unethical marketing around the world. (Klein 2001: xx)

Klein's conceptualization of activists as spiders on a global web provides the beginnings of a practical strategy with which to approach the confusing immateriality of modern capitalism. It is in keeping with Dyer-Witheford's call for oppositional groups to match the nomadic flows enjoyed by corporations due to their own 'global-webs' of capital (Dyer-Witheford 1999: 143). The need for a counter-colonization of the global web is now an increasingly common call amongst radical thinkers. To those previously cited can be added Hardt and Negri (2000) whose basic premise of the need for opposition to a new global corporate *Empire*, relies heavily upon the belief that its web of capital flows and commodity circulation needs counter-populating with flows of struggle from different communities within "the global multitude" (Hardt & Negri, 2000: 46). It is interesting to note the similarity of their language with the previously cited fictional comparison of the spread of corporate values with biological viruses: 'Rather than thinking of the struggles as relating to one another like links in a chain, it might be better to conceive of them as communicating like a virus that modulates its form to find in each context an adequate host.' (Hardt & Negri, 2000: 51). Their call for the "counter-populating" of "the global multitude" has been answered by various groups of hacktivists seeking to develop new Web-based tactics to better confront the new online forms of capital.

## SEMIOLOGICAL GUERILLA WARFARE

*In times of constant effervescence, certain stimulating imperti-nences are required.* (Maffesoli 1996: 7)

*In technological forms of life, not just resistance but also power is non-linear. Power itself is no longer primarily pedagogical or narrative but instead, itself performative. 'Nation' now works less through 'narrative' or 'pedagogy' but through the performativity of information and communication. Power works less through the linearity and the reflective argument of discourse or ideology than through the immediacy of information, of communications.* (Lash 2002: 25)

Copyright © 2003, Idea Group Inc. Copying or distributing in print or electronic forms without written permission of Idea Group Inc. is prohibited.

According to Lash, the performativity of information is the dominant factor in the spread of global communication systems. Put in simpler terms, Lash's argument can be seen as a variation upon McLuhan's aphorism: *the medium is the message.* The ubiquitous immanence of information and communication technologies means that all social meaning becomes disproportionately mediated through the prism of immediate, functional data rather than the nuanced and less time-obsessed nature of more reflective and analytical thought. In a much earlier analysis of mass communication systems, Eco (1967) reinforces this analysis by arguing that there is little room for an optimistic reinterpretation of the innately deterministic implication of McLuhan's famous phrase. Eco recognizes that the meanings derived from communicated messages are filtered through the social codes we bring to them, but then argues such room for reinterpretation of the dominant code behind mass communication systems is extremely limited:

> There exists an extremely powerful instrument that none of us will ever manage to regulate; there exist means of communication that, unlike means of production, are not controllable either by private will or by the community. In confronting them, all of us from the head of CBS to the president of the United States, from Martin Heidegger to the poorest fellah of the Nile delta, all of us are the proletariat. (Eco, 1967: 141)

Confronted by this situation, Eco distinguishes between a strategic and tactical approach. The former aims to fill the existing channels of communication with radically like-minded people who can seek to fill those channels with liberating opinions and information. As the above quotation illustrates, however, the likelihood of success is limited because as Eco puts it, the "means of communication ... are not controllable by private will or by the community". He suggests that such an approach may achieve good short term political or economic results: 'but I begin to fear it produces very skimpy results for anyone hoping to restore to human beings a certain freedom in the face of the total phenomenon of Communication.' (Eco 1967: 142)

New online activist groups have taken Eco at his word and developed various tactical semiological performances and events designed to shock people from the passivity of total communication's regime in ways that belatedly promise to fulfill his call for a new form of "semiological guerilla warfare".

Illustration is provided by the way in which traditional forms of civil disobedience such as peaceful sit-ins have been transformed, in cyberspace, into new forms of *electronic civil disobedience*. In 1998, for example, the hacktivist group the *Electronic Disturbance Theatre* (EDT) coordinated a series of Web sit-ins in support of the Mexican anti-government group, the Zapatistas. This incident was perhaps most noticeable for its use of an automated piece of software revealingly

Copyright © 2003, Idea Group Inc. Copying or distributing in print or electronic forms without written permission of Idea Group Inc. is prohibited.

called *FloodNet*. Once downloaded on to an individual's computer automatically, this piece of software connects the surfer to a pre-selected Web site. Every seven seconds the selected site's reload button is automatically activated. If thousands of people use FloodNet on the same day, the combined effect of such a large number of activists will disrupt the operations of a particular site. Similar techniques were used by another group, ®™ark, in the e-toy campaign of 1999. This was a hacktivist response to a commercial company's attempt to use the courts to remove an art collective's Web site domain name because they felt it was too similar to their own[6]. In what was described as the "Brent Spar of e-commerce"[7], a combination of Internet and media public relations stunts were used to force an eventual *volte-face* by the company, greatly aided by the 70 percent decline in the company's NASDAQ stock value that coincided with these actions.

# ®™ARK AND THE REVERSE ENGINEERING OF THE CORPORATE MODEL

*... a future communications guerilla warfare – a manifestation complementary to the manifestations of Technological Communication, the constant correction of perspectives, the checking of codes, the ever renewed interpretations of mass messages. The universe of Technological Communication would then be patrolled by groups of communications guerillas, who would restore a critical dimension to passive reception. The threat that the "the medium is the message" could then become, for both medium and message, the return to individual responsibility. To the anonymous divinity of Technological Communication our answer could be: "Not Thy, but our will be done."* (Eco 1967: 144)

To oppose "the total phenomenon of Communication", and because of capitalism's ability to co-opt and submerge oppositional forces premised upon more strategic approaches. Eco proposes the above general outline of a tactical approach which as proved to be an extremely prescient description of what we shall now see are the actual tactics adopted by such new radical online groups as ®™ark. This is an on-line activist group that provides some of the best examples of such a political cause being translated into practical action in the form of ®™ark projects. These are based upon the four "keys" of worker, sponsor, product, and idea: '®™ark is a system of workers, ideas, and money whose function is to encourage the intelligent sabotage of mass-produced items ... ®™ark is essentially a match-

Copyright © 2003, Idea Group Inc. Copying or distributing in print or electronic forms without written permission of Idea Group Inc. is prohibited.

maker and bank, helping groups or individuals fund sabotage projects' ("A System for Change" ®™ark Undated Website paper)

The Group's web site provides numerous examples of past and present projects that it has supported in a form deliberately modelled upon that of the financial mutual fund system. Notable examples include the setting up of a Web site entitled *Voteauction.com* that aimed and succeeded in attracting press attention for the way in which it purported to buy votes from people. The project successfully provoked a series of media reports that felt obliged to comment upon the widespread perception that democracy had already been more substantially compromised by the way in which large corporations already effectively "buy" votes through their lobbying power. Other projects involved the setting up of a fake W.T.O. site in order to satirize the organization and its G.A.T.T. agreements. A final example is provided by the *Barbie Liberation Organisation* who, as a mixed group of activists and military veterans switched the voice boxes in 300 Talking Barbie dolls and Talking G.I. Joe dolls during the 1989 Christmas period with the goal of highlighting and correcting the problem of gender-based stereotyping in children's toys.

Their imitative *modus operandi* mimics the capitalist investment process in order to better subvert it. They act as a perverse form of commercial clearing and investment house for projects they think will provide a "good return" in terms of human capital. Their close satirical imitation of capitalist structures and practices fits well with Eco's desire to see "a manifestation complementary to the manifestations of Technological Communication whilst their rationale is expressed in terms resonant of Eco's previously cited call "to restore to human beings a certain freedom in the face of the total phenomenon of Communication": '®™ark ... is an ark for our humanness through the deluge of ® and ™ ... an attempt to make our environment more palatable, more reflective of us, and generally more human.' (Ibid, 1997)

To help conceptualize this pursuit of more humane aims, the group borrows the term "curation" from the world of art. Artists are seen as an important cultural reservoir of non-commodified values and the aim is to spread such values out into the broader society. "Curates" is used as a synonym for "influences" and describes, in a fashion very much in keeping with Hardt and Negri's (2000) notion of biopolitics, the way in which daily life becomes inseparable from the formative influence of advertisements or consumer brands. Thus in ®™ark's view, citizens are only valued as "input mechanisms" for consumer values and all the objects one confronts in social life have been "curated" to facilitate this input: 'Even those curated objects which seem to encourage creation only encourage such creation as leads without delay to consumption, either one's own (games, art technologies,

Copyright © 2003, Idea Group Inc. Copying or distributing in print or electronic forms without written permission of Idea Group Inc. is prohibited.

etc.) or that of others (work).' (®™ark Curation Website paper 1998) In keeping with hacktivism's general philosophy of reverse engineering (originally inherited from hacking [see Taylor in Wall 2001]), ®™ark sets itself up in direct opposition to this tendency to define people solely as consumerist input mechanisms: it appeals to citizens as creative output mechanisms. It is here that the *performance* element of hacktivist activities comes to the fore. ®™ark recognizes the combined narcotic effects of both the media and the pervasive social environment of consumption that it reinforces. The group thus seeks to use various media performances targeted at citizens-as-consumers in order to jolt them out of their uncritical contentment based upon the unthinking consumption of commodities:

> It was once the case that advertising appealed to our insecurities and miseries, and tried to exacerbate existential troubles in order to offer costly solutions … but these methods have been swallowed by the very fear they generated. Just as repression has wisely given way to choicelessness, exacerbation has given way to anaesthetic. Content- ment, though more expensive than terror, is in the long run cheaper, since the price for contentment can be set: as consumption. Ultimately, contentment pays for itself. (®™ark Globalization and Global resistance Website paper)

®™ark sees a wellspring of potential subversion within growing levels of social discontent. Their projects seek to build upon rising levels of disaffection where people are becoming much more irrational, unpredictable and creative. ®™ark reappropriates the imminent communicational and informational performance previously identified by Lash (2002) and translates it into a superficially similar but fundamentally subversive format. The Electronic Disturbance Theatre (EDT) to which we now turn adopts the tactic of performance in an even more radical and semiologically upsetting *politics of magic realism*.

## THE ELECTRONIC DISTURBANCE THEATRE

*The Zapatistas use the politics of a magical realism that allows them to create these spaces of invention, intervention, and to allow the worldwide networks to witness the struggle they face on daily. It was the acceptance of digital space by the Zapatistas in 12 days that created the very heart of this magical realism as information war. It was this extraordinary understanding of electronic culture that allowed the Zapatistas on 1 January, 1994, one minute after midnight just as a Free Trade Agreement between Canada, U.S.A,*

Copyright © 2003, Idea Group Inc. Copying or distributing in print or electronic forms without written permission of Idea Group Inc. is prohibited.

*and Mexico (NAFTA) went into effect - to jump into the electronic fabric, so to speak, faster than the speed of light. Within minutes, people around the world had received e-mails from the first declaration from the Lacandona Jungle. The next day the autonomous Zapatista zones appeared all over the Internet. The New York Times considered it the first post-modern revolution. The American intelligence community called it the first act of social net war. Remember, that this social net war was based on the simple use of e-mail and nothing more ... gestures can be very simple and yet create deep changes in the structures of the command and control societies that neo-liberalism agenda, like NAFTA, represent.*
(Ricardo Dominguez of the EDT in Fusco, 1999)

The Electronic Disturbance Theatre has been at the forefront of developing both specific semiological guerilla tactics and an over-arching tactical ethos. The specific tactics have so far taken the form of mass online actions reinforced by the use of symbolic/performative semantics to create the groundswell of online empathy required to maintain the neo-tribe. In the particular example of *FloodNet*, the website of Mexico's President Zedillo was overwhelmed by the coordinated efforts of physically disparate activists. The failure of his site due to such collected action begins to hint at the immaterial forms public space in cyberspace can assume. The FloodNet action vividly illustrates the effectiveness of mass political participation in the virtual realm. The lack of physical space in which to meet is compensated for by the binding empathy created by the positive fall-out from the disturbance effects of online actions: 'The FloodNet gesture allows the social flow of command and control to be seen directly – the communities themselves can see the flow of power in a highly transparent manner.' (Dominguez in Fusco, 1999) The questioning of this flow of power to provide greater transparency is complemented by actions designed to make an additional political point through such artistic expressions as the creative use of 404 files. These files are the standard Web page replies that a user receives when information they have sought is not available from the server:

We ask President Zedillo's server or the Pentagon's web server 'Where is human rights in your server?" The server then responds "Human rights not found on this server" ... This use of the "not found" system ... is a well-known gesture among the net art communities. EDT just re-focused the 404 function towards a political gesture. (Dominguez in Fusco, 1999)

Copyright © 2003, Idea Group Inc. Copying or distributing in print or electronic forms without written permission of Idea Group Inc. is prohibited.

404 file art and the above description by Dominguez of *digital zapatismo* vividly illustrates how Maffesoli's (1996) "empathetic" and effectual neo-tribes are manifested in practice. Despite the essentially immaterial nature of the effectual environment, the "extraordinary understanding of electronic culture" facilitated the emergence of a global neo-tribe of like-minded radicals. This empathy is both a cause and a consequence of specific mass online actions. It also somewhat complicates Hardt and Negri's (2000) assertion that horizontal chains of political action have been supplanted in the era of global communication by "vertical media events". The experience of *digital zapatismo* implies that global awareness of site-specific struggles results from the pre-existence of horizontally nurtured links that then spring vertically upwards into the gaze of the global media. The circulation of struggle thus occurs:

> via a strange chaos moving horizontally, non-linearly, and over many sub-networks. Rather than operating through a central command structure in which information filters down from the top in vertical and linear manner information about Zapatistas on the Internet has moved laterally from node to node. (Dominguez in Fusco, 1999)

It is perhaps inevitable that the media tends to privilege the technical vehicles of the protests over their political and social content, but for Dominguez, the actual form the protests take is the least important aspect of a larger and more significant three-act performance. The first act involves stating what is going to happen and its political purpose, the second is the act itself, and the third is the subsequent dialogue and discussion that creates what Dominguez calls a "social drama": 'A virtual plaza, a digital situation, is thus generated in which we all gather and have an encounter, or an Encuentro, as the Zapatistas would say – about the nature of neo-liberalism in the real world and in cyberspace.' (Dominguez in Fusco, 1999) *Digital zapatismo* has added an additional element to such social drama by using periods of tactical silence where, literally in Mexico and metaphorically elsewhere, the activists retreat back into the jungle for a period of calm reflection - the effect of which is heightened by its contrast with the media's need for the constant noise of news.

# A DOT.COMMUNIST MANIFESTO - RECLAIMING THE AGORA

In this chapter's discussion of *hacktivism*, we have concentrated predominantly upon the "ism" part of the word, which refers to political activism that

Copyright © 2003, Idea Group Inc. Copying or distributing in print or electronic forms without written permission of Idea Group Inc. is prohibited.

motivates acts and events. However, the "hack" part of the word relates to the earlier practice of hacking. In my previous detailed exploration of this phenomenon (Taylor, 1999) a fundamental aspect of the true "hack" was its innate desire to re-engineer or reverse the original and primary purpose of an artifact or system. Hacktivism has kept faith with this quality; see Dominguez's over-arching rationale below for not only the performance element of hacktivism, but also its tactic of re-appropriating and re-colonizing the space that e-commerce has so far claimed as its own:

> The idea of a virtual republic in Western Civilization can be traced back to Plato, and is connected to the functions of public space. The Republic incorporated the central concept of the Agora. The Agora was the area for those who were entitled to engage in rational discourse of Logos, and to articulate social policy as the Law, and thus contribute to the evolution of Athenian democracy. Of course those who did speak were, for the most part, male, slave-owning and ship-owning merchants, those that represented the base of Athenian power. We can call them *Dromos*: those who belong to the societies of speed. Speed and the Virtual Republic are the primary nodes of Athenian democracy – not much different than today. The Agora was constantly being disturbed by Demos, what we would call those who demonstrate or who move into the Agora and make gestures. Later on, with the rise of Catholicism – Demos would be transposed into Demons, those representatives of the lower depths. Demos did not necessarily use the rational speech of the Agora, they did not have access to it; instead, they used symbolic speech or a somatic poesis - Nomos. In the Agora, rational speech is known as Logos. The Demos gesture is Nomos, the metaphorical language that points to invisibility, that points to the gaps in the Agora. The Agora is thus disturbed; the rational processes of its codes are disrupted, the power of speed was blocked. EDT alludes to this history of Demos as it intervenes with Nomos. The Zapatista FloodNet injects bodies as Nomos into digital space, a critical mass of gestures as blockage. What we also add to the equation is the power of speed is now leveraged by Demos via the networks. Thus Demos_qua_Dromos create the space for a new type of social drama to take place. Remember in Ancient Greece, those who were in power and who had slaves and commerce, were the ones who had the fastest ships. EDT utilizes these elements to create drama and movement by empowering contemporary groups of Demos with the speed of Dromos – without asking societies of command and control for the right to do so. We enter the Agora with

Copyright © 2003, Idea Group Inc. Copying or distributing in print or electronic forms without written permission of Idea Group Inc. is prohibited.

the metaphorical gestures of Nomos and squat on high-speed lanes of the new Virtual Republic – this creates a digital platform or situation for a techno-political drama that reflects the real condition of the world beyond code. This disturbs the Virtual Republic that is accustomed to the properties of Logos, the ownership of property, copyright, and all the different strategies in which they are attempting enclosure of the Internet. (Dominguez in Fusco, 1999)

Dominguez's imaginative reinterpretation of the demos is empowered by the speed of the system that hacktivist groups seek to redirect. The new social dramas that result are thus given additional bite, because they achieve their effects from within the system rather than from a "pure" intellectual distance. The danger of replicating through online actions the abstract rationalism that is being protested against is avoided because hacktivists remain mindful of "the real condition of the world beyond code". In keeping with the central role of performance and social dramas to the anti-Globalization tactics of indigenous groups such as the Zapatistas, hacktivists have reinvented Nomos for the agora of the $21^{st}$ century. Through their hands-on activism, they have reclaimed the spirit of magic realism from its imprisonment in university literature departments and they confront head-on Logos in its new guise of the logo.

# CONCLUSION

*She drives past Clare's Agricultural Division headquarters at least three times a week. The town cannot hold a corn boil without its corporate sponsor. The company cuts every other check, writes the headlines, and sings the school fight song. It plays the organ at every wedding and packs the rice that rains down on the departing honeymooners. It staffs the hospital and funds the ultrasound sweep of uterine seas where Lacewood's next of kin lie grey and ghostly, asleep in the deep.* (Power 1998: 6)

*We should be done once and for all with the search for an outside, a standpoint that imagines a purity for our politics. It is better both theoretically and practically to enter the terrain of Empire and confront its homogenizing and heterogenizing flows in all their complexity, grounding our analysis in the power of the global multitude.* (Hardt & Negri, 2000: 46)

Copyright © 2003, Idea Group Inc. Copying or distributing in print or electronic forms without written permission of Idea Group Inc. is prohibited.

*Gain* (1998) is a novel about the history of a multinational, Clare Soap and Chemical Company. In the above excerpt, the protagonist is suffering from a cancer caused by the ubiquitous firm's pollution in her local community. The author uses her individual predicament to represent the wider social impact of increased blurring that has taken place between commerce and society, a blurring that is likely to increase in the age of e-commerce. Thus the key significance of the easily-reinsertable, decontextualized quality of e-commerce previously described by Ullman (1997) is the resultant ease with which commodity values have both pervasive and invasively destructive effects upon the cultural fabric: in Hardt and Negri's (2000) terms, they exert a *biopolitical power*. Their subsequent call to "enter the terrain of Empire" has been met by online groups who, while maintaining their groundedness in physical social reality, have nevertheless, enthusiastically sought to build new immaterial social spaces square in the middle of the Empire's territory.

To oppose the Empire, the EDT reverse-engineer the functional performativity of the binary-based global communication regime. It does this by refocusing its instrumental emphasis upon immediacy with spontaneous Web-based actions, and also by problematizing it through the explicit contrast created by the satirical, magic realist quality of those actions. The events it creates resonate beyond their immediate disturbance to provoke the viewer/reader into deeper reflection about their significance. They throw into sharp relief the dominant logic that, as we have seen throughout this chapter, relies upon the immanent immediacy of rapidly circulating communication. I have previously pointed out that this reflective process is reinforced by the creative silences that Digital Zapatismo in particular has utilized immediately after several high profile events. In the reflective moment afforded by a conclusion, perhaps the issue to contemplate is the extent to which, after the widely-perceived failure of International Socialism, there may be a new force ready and willing to confront the smooth advance of friction-free e-capital. As quoted near the beginning of this chapter, Marx claimed that as a by-product of capitalism's global reach, "there arises a world literature." Magic realism may yet prove to be that literature, to paraphrase Marx: 'A spectre is haunting the globe – the spectre of hacktivism'.

# ACKNOWLEDGMENTS

My initial co-operation with Ricardo Dominguez was made possible by funding from the UK's Economic & Social Research Council (ESRC) and its seminar series on *Living in the Matrix: Immateriality in theory and practice*. Quotations cited as (Fusco, 1999) were taken from an interview entitled, "Performance Art in a Digital Age: A Conversation with Ricardo Dominguez" that

Copyright © 2003, Idea Group Inc. Copying or distributing in print or electronic forms without written permission of Idea Group Inc. is prohibited.

took place on Thursday 25 November 1999, at the Institute of International Visual Arts. The interview was heavily edited by Coco Fusco and transcribed by InIVA staff. It was republished in Centrodearte.com and Latinarte.com.

# REFERENCES

Berman, M. (1983). *All that is Solid Melts into Air: The Experience of Modernity.* London: Verso.

De Certeau, M. (1988). *The Practice of Everyday Life.* Berkeley: University of California Press.

Deleuze, G. (1989). *Cinema 2: The Time-Image.* Minneapolis: University of Minnesota Press.

Dyer-Witheford, N. (1999). *Cyber-Marx: Cycles and Circuits of Struggle in High-Technology Capitalism.* Chicago: University of Illinois Press.

Eco, U. (1987). *Travels in Hyperreality.* London: Picador.

Frank, T. (2001). *One Market Under God: Extreme Capitalism, Market Populism and the End of Economic Democracy.* London: Secker and Warburg.

Fusco, C. (1999) *Performance Art in a Digital Age: A Conversation with Ricardo Dominguez.* Unpublished paper.

Greider, W. (1997) *One World, Ready or Not.* New York: Touchstone.

Grether, R. (2000). How the e-toy campaign was won. *Telepolis* (Jan 2000). Available at: http://www.heise.de/tp/english/inhalt/te/5843/1.html.

Hardt, M. & Negri, A. (2000). *Empire.* Cambridge, MA: Harvard University Press.

Klein, N. (2001). *No Logo.* London: Flamingo.

Lash, S. (2002). *Critique of Information.* London: Sage.

Lasn, K. (2000) *Culture Jam: How To Reverse America's Suicidal Consumer Binge – and Why We Must.* New York: HarperCollins.

Leadbetter, C. (2000). *Living on Thin Air: The New Economy.* London: Penguin Books.

Lefebvre, H. (1991). *The Production of Space.* Oxford: Blackwell.

Levitt, T. (1983) The globalization of markets. *Harvard Business Review*, May-June, cited in Klein, 2000.

Maffesoli, M. (1996) *The Time of the Tribes: The Decline of Individualism in Mass Society.* Sage: London.

Marx & Engels in Tucker (ed.) (1978) *The Marx-Engels Reader.* New York: W.W. Norton.

Copyright © 2003, Idea Group Inc. Copying or distributing in print or electronic forms without written permission of Idea Group Inc. is prohibited.

Powers, R. (1998) *Gain*. London: William Heinemann.
Schlosser, E. (2001). *Fast Food Nation: What the All-American Meal is doing to the World*. London: Penguin Books.
Stephenson, N. (1992) *Snowcrash*. New York: Bantam Spectre.
Taylor, P.A. (1999) *Hackers: Crime in the Digital Sublime*. London: Routledge.
Taylor, P.A. (2001) Hacktivists: in search of lost ethics? In Wall, D. (ed.) 2001: *Crime and the Internet*. London: Routledge.
Ullman, E. (1997). *Close to the Machine: Technophilia and its Discontents*. San Francisco: City Lights Books.


**Web-based sources** (checked June 2002)
RTMark papers

"A System for Change"
"Curation"
"Globalization and global resistance"

All the above RTMark papers can be found at their Web site: http://www.rtmark.com.

# ENDNOTES

1 The historical and perennial significance of this analysis by Marx is explored in detail in Marshall Berman's *All That is Solid Melts Into Air* (1983).
2 Cited in Greider, 1997: 24.
3 The Guardian, Feb 7 , 2001.
4 Ibid., August 2, 2001.
5 The *Daz-Doorstep challenge* refers to a soap-powder advertisement on British TV, whereby a celebrity challenges a series of housewives, on their doorsteps, to use the Daz product.
6 For a full account. see Grether (2000).
7 See the RTMark press release, available at: http://www.rtmark.com/etoyprtriumph.html.

Copyright © 2003, Idea Group Inc. Copying or distributing in print or electronic forms without written permission of Idea Group Inc. is prohibited.

## Chapter II

# The Economic and Social Impact of Electronic Commerce in Developing Countries

Roberto Vinaja
University of Texas, Pan American, USA

## ABSTRACT

*The chapter addresses the potential benefits of Electronic Commerce to developing countries. Electronic commerce has many potential benefits for developing countries (DC). In this chapter, the author describes the potential benefits of Electronic Commerce for developing nations and he provides case examples that illustrate this trend. The widespread adoption of electronic commerce is especially important for developing countries.*

Copyright © 2003, Idea Group Inc. Copying or distributing in print or electronic forms without written permission of Idea Group Inc. is prohibited.

# INTRODUCTION

Electronic commerce has many potential benefits for developing countries (DC). In this chapter we will describe the potential benefits of Electronic Commerce for developing nations and provide case examples that illustrate this trend. The widespread adoption of electronic commerce is especially important for developing countries. The benefits for developing countries range from social to economic. Some of the benefits include: improvement of international coordination, an open economy promoting competitions and diffusion of key technologies, efficient social and infrastructure services, a competitive communication sector, and increased buyer productivity.

The impacts of electronic commerce in a developing country can be helpful rather than detrimental. Electronic commerce has the potential to tie developing countries into the rest of the world so they are no longer considered outsiders. For example, electronic commerce can enable more people to access products and services that once were not available. Another benefit is that electronic commerce stores are available 24 hours a day, 7 days a week. As the infrastructure for electronic commerce keeps growing, services that were not offered in the past become available. Many of these benefits have not been proven yet, but the technology is now available, and developing countries are looking forward to these benefits. The high cost of technology may still be detrimental in many developing countries; however, the constant innovation of software and hardware will hopefully reduce these costs.

Consumers in developing countries can benefit from electronic commerce because they can buy products that could only be found in major cosmopolitan cities. Electronic commerce is closing the gap between those countries that have wide availability of products and those with limited availability. The basic purpose of electronic commerce is to provide goods and services to consumers who do not live close to the physical location of the product or service and would otherwise have a hard time acquiring these products and services.

Society and consumers alike have only begun to enjoy the benefits of electronic commerce. Since new developments are made on a continuous basis, it will eventually affect every individual. Some of the benefits enjoyed by society and consumers, for example, are ease of transaction, comparability of products, quick delivery and the ability to make any type of transaction at any given time of day.

Electronic commerce facilitates delivery of public and social services, such as healthcare, education, and distribution of government social services at a reduced cost, improving the quality of care and living in these communities. For example, health care services can reach patients in rural areas (Turban et al., 2000).

Copyright © 2003, Idea Group Inc. Copying or distributing in print or electronic forms without written permission of Idea Group Inc. is prohibited.

Another benefits is the fast dissemination of information, information is distributed in a matter of seconds, instead of several months. Electronic commerce can help people become better educated and better informed. Many educational opportunities are becoming available to developing countries. For example, the availability of virtual universities provides the opportunity to learn and earn college degrees. In addition, many developing-country universities are focusing on curricula that might contribute more directly to economic growth, and network connections for administrators, professors, and students will be increasingly important.

The communications and information delivery capability of the Internet can benefit all sectors of society. The areas of education, health, social policy, commerce and trade, government, agriculture, communications, and science and technology could benefit from the improved access to information provided by the Internet (Sadwosky, 1996).

Access to information affects political democratization efforts at the global level as well as within nations. There seems to be a connection between the free flow of information and movement toward democratization. This fact has been observed in a number of countries recently (Hay et al., 2000).

# NEW DEVELOPMENTS

Several organizations such as the ITU (International Telecommunication Union), the GII (Global Information Infrastructure), the NII (National Information Infrastructure), the OECD (Organization for Economic Cooperation and Development), and the EU (European Union), are striving to develop standards and policies to promote global Electronic commerce. (Hay et al., 2000). The ITU is a worldwide organization, where 189 member states and some 570 sector members representing public and private companies and organizations with an interest in telecommunications cooperate for the development of telecommunications and the harmonization of national telecommunication policies.

In 1998, the International Telecommunications Union, in conjunction with The World Trade Center in Geneva and the World Internet Service Key launched the "Electronic Commerce for Developing Countries" (EC-DC) project (Ntoko, 1999). The project Electronic Commerce for Developing Countries (EC-DC) is an activity of the ITU Telecommunication Development Bureau (BDT). EC-DC assists developing countries in the use of electronic commerce by addressing the technology, policies and strategy issues related to electronic commerce. It provides a framework for neutral and non-exclusive partnerships with industry thereby creating the environment for cost-effective solutions to the benefit of developing

Copyright © 2003, Idea Group Inc. Copying or distributing in print or electronic forms without written permission of Idea Group Inc. is prohibited.

countries. It aims at enabling developing countries to use existing infrastructures and services to participate in electronic commerce. It also seeks to facilitate the transfer of electronic commerce technology, increase public awareness and stimulate the planning and deployment of the telecommunication infrastructure (Ntoko, 2000).

The purpose of this project is to expand electronic commerce in developing countries and to help in the construction of electronic commerce infrastructure and implementation of electronic commerce solutions (Goh, 2001). The EC-DC project will assist the ITU in expanding electronic commerce in the Developing Countries by using the World Trade Center network and its global resources of more than 300 centers in more than 100 countries. The International Telecommunication Union's web project "Electronic Commerce for Developing Countries" is being deployed in more than 100 countries. It is a massive project under which participating countries can benefit from first-class security, trust and services for e-business transactions under affordable conditions.

The EC-DC project addresses some of the challenges and opportunities faced by developing countries in the application of new technologies. The ITU is assisting developing countries to acquire and benefit from electronic commerce technologies through a program focused on concrete deliverables. The EC-DC project has identified the key benefits of electronic commerce for developing countries:

- *Economic Development:* "micro and small businesses can begin to trade at internationally acceptable price levels and bypass the system of exploitation of their products for minimal return" (Ochienghs, 1998). The companies are able to trade at internationally acceptable price levels and among and across many borders. Benefits in the tourism, travel, arts, sale of locally produced goods, service industry and the banking sector are greatly seen because of the ability of EC to reduce the cost of processing orders and payments in the global marketplace. This in turn contributes to an "economic upliftment" (ITU, 2001a).

- *Infrastructure Development:* EC will stimulate demand for Internet connection infrastructure and encourage the creation of commerce (Gagné, 2001). As a result, policies in the banking and information and communication technology sectors will bring together public and private "communities" to support infrastructure development. In order to provide immediate benefits and increase the chances of sustainability of the project, the initial focus of the EC-DC project has been on businesses that already have an export market. The project also enables the transfer of electronic commerce technologies through the development of human resources necessary for providing electronic commerce services and maintaining the infrastructure (Gagné, 2001).

- *Regional and multi-national cooperation:* The improvements in the infra-

Copyright © 2003, Idea Group Inc. Copying or distributing in print or electronic forms without written permission of Idea Group Inc. is prohibited.

structure will facilitate communication between business and consumers (Ochienghs, 1998). EC "increases the collaboration between the various sectors of government, banking, business and information technology of the country" (ITU, 2001b). One of the social benefits that electronic commerce provides to developing countries is the partnership among different countries. More jobs are created as demands for products rise due to the large exposure electronic commerce has allowed.

# CASES

## Panama
Panama has implemented many initiatives for EC (National Law Center, 2000). Panama's tourism has benefit from this by putting information of the best places to visit in their country on the web.

Travel agencies provide information at their web sites that are virtually available to any person in the world interested in visiting Panama. There are plenty of benefits for organizations, such as saving on the cost of brochures, advertising in international newspapers or magazines, and on methods of payments that they have to do through their bank (U.S. Department of Commerce, 2000).

## Tunisia
For example, today electronic commerce is a reality in Tunisia, with pilot projects selling Tunisian products in all countries of the world and a bill for electronic exchanges and commerce that was presented to the Chamber of Deputies. The object of the year 2000 was to generalize the use of this new mode of commerce in Tunisia and create public and private online services allowing Tunisian citizens to take full advantage of electronic commerce (ISOC, 2001).

## Malaysia
In 1997, government officials in Malaysia noticed that the rapid diffusion of the Internet throughout the world had accelerated the introduction of electronic commerce. The government envisioned a profound structural change in the economy of the country and a significant impact on international trade. Malaysia's electronic commerce expansion is also partially credited to the growth in PC purchase and use in that nation, the actual hardware and is somewhat the backbone of electronic commerce. The growth of PCs in the nation provides another source of electronic commerce, C2B and C2C. The new electronic commerce markets are

Copyright © 2003, Idea Group Inc. Copying or distributing in print or electronic forms without written permission of Idea Group Inc. is prohibited.

giving Malaysia more opportunities from within the country and not just from foreign nations. The significance of PC growth, in relation to electronic commerce, is that Malaysian citizens will also have access to the global markets to be able to buy products and services at a much cheaper cost. So, electronic commerce is more than just Malaysia's ability to supply global, potential buyers with products; it is an overall impact on Malaysia's economy, thanks to the various implementations of electronic commerce (Cordelia, 1999). In conclusion, Malaysia has benefited from implementing electronic commerce into their society but they must aspire to improving much more. "It is about transforming our current economy, which is dependent on commodities and contract manufacturing, into a different economy that uses Information and Communications Technology (ICT) to improve manufacturing processes, reduce manpower needs, lower costs of production, find better and more profitable uses for commodities and eventually improve the quality of life of Malaysians." (Sivapalan, 2001)

## Africa

An example how the electronic commerce has provided a lot of different benefits for a developing country like Africa. "The African continent has the least developed telecommunication network in the world" (Coeur, 1997). One of those benefits is for consumers. Because of the size of Africa, electronic commerce provides most of its citizens with availability to countless products from the Internet. Without the Internet, many of Africa's citizens would have a very hard time finding things that are not available in their small town stores. Even if they were to find the items that they were looking for, they would have to travel very long distances to get them.

The Prime Minister of Mozambique, H.E. Mr. Pascoal Mocumbi, inaugurated the first telemedicine link of Mozambique in 1999. It is one of the firsts in Africa. The central hospitals of Beira and Maputo are able to making use of standard low-cost teleradiology equipment which provides support for the exchange and visualization of images including radiographs as well as for transmitting laboratory results or for communication (verbal or written messages). This example shows how telemedicine can help overcome some of the serious shortages in health care services in developing countries (Androuchko, 1998).

## Nicaragua

Some electronic stores sell handcrafted goods from developing countries, buyers get beautiful, unique, hand-crafted goods, and their purchase preserves native arts, lets villagers stay at home rather than migrating to city factories, and

Copyright © 2003, Idea Group Inc. Copying or distributing in print or electronic forms without written permission of Idea Group Inc. is prohibited.

reduces poverty. Money from selling their wares has given villagers the means to build new houses and workshops (Richman, 1999).

## Mexico

For Mexico and other Latin American countries, tourism is of great economical impact; therefore, a large number of web sites are dedicated to support, promote national and international tourism. For any developing country electronic commerce is a great way to enter into foreign markets and compete in the rapidly growing global economy. In the case of Mexico, the recent implementation of NAFTA has led them to acquire some of the technological advancements from the U.S. and Canada. This includes electronic commerce applications. According to John F. Smith, the Chief Executive Officer and President of General Motors Corporation, global communication links also make transportation faster and less expensive than anybody would have imagined just a few years ago, which makes it much simpler to conduct business. The benefits that electronic commerce brings to Mexico are much the same as those in other countries, but since Mexico is a developing country, it provides them even greater benefits. Electronic commerce reduces inventory and overhead. In addition, distributors are able to customize the products and services to better fit the needs of consumers, suppliers, and employees. Mexican consumers are also enjoying access to services and products manufactured in foreign countries that were not accessible to them before. Many Mexican nationals are now earning college credits from foreign universities who offer on-line courses while at the comfort of their homes. Countries like Mexico are coming along quickly and taking advantage of the benefits of electronic commerce.

## Sri Lanka

Electronic commerce has opened many doors for the society in Sri Lanka; one of the most important is an increase of their productivity and competitiveness in international trade. Sri Lanka's economy is based on exports of tea, textiles, agricultural produce, graphite and manufacturing goods. At the beginning of 1997 and 1998, when Sri Lanka started to develop IT, "the electronic commerce trend was about scoring hits at websites and selling products at very low prices, not taking into consideration that selling at a loss doesn't translate into profits" (Sri Lanka Telecom, 2001a). Also to develop a better communication within the supply chain, Sri Lanka formed a National EDI Committee in 1995 to provide network services so businesses can improve computer-to-computer direct transfer of standard business documents. A great example of this technology innovation is the Commercial Bank. The strategy of this bank is to expand their retail presence by offering ATMs and anywhere, anytime banking". The major benefit from this initiative is that

Copyright © 2003, Idea Group Inc. Copying or distributing in print or electronic forms without written permission of Idea Group Inc. is prohibited.

they "offer a customized solution, which would suit each individual need (B2B or B2C)" (Ariyadasa, 2000).

The benefits have not only reflected in the increment of their productivity, but also in the implementation of education. The National Institute of Education conducted a real time distance education program in Sri Lanka. This program is carrying out on ISDN using video telephones. The director of the General Education Institute mentioned "this effort will definitely take Sri Lanka to a new era of education, a borderless education program" (Sri Lanka Telecomm 2001b). Also the Lanka Experimental Academic and Research Network has provided email facilities to more than 50 sites including universities and research institutions via dial up lines (providing oversees connectivity).

## FUTURE TRENDS

Rapid expansion of the Internet holds substantial promise for developing nations. They can benefit greatly from the Internet's communication and information delivery capabilities to help meet these needs. (Hay et al., 2000) Developing countries have much to gain from that revolution in communication and information access.

Electronic commerce will give developing countries an 'increase in jobs, wealth and health. With the creation of new jobs the economy improves. EC will also provide help, services and jobs, the reduction of unemployment and the growth of GDP, this will also result in an improvement of the living conditions of the population.' (African Development Forum, 2000)

EC will have potential economic advantages in travel, tourism, sale of locally produced goods, and the banking industry, because it lowers the cost of processing orders and payments and is accessible to the global market. Electronic commerce allows small businesses to sell their products from anywhere in the world. At the long run, producers will be able to provide better customer service and will enable people in Third World Countries and rural areas to enjoy products and services that otherwise are not available to them (Turban et al., 2000).

## CONCLUSION

Many governments across the regions have recognized the tremendous opportunities that the Internet offers, "priority of governments to support the development of internet-access infrastructure, to stimulate growth and encourage businesses and individuals to go on-line (Hillebrand, 1999). The changes electronic

Copyright © 2003, Idea Group Inc. Copying or distributing in print or electronic forms without written permission of Idea Group Inc. is prohibited.

commerce will bring are far-reaching. They require new frameworks for doing business and a re-examination of government policies relating to commerce and skills. Governments of the developing countries have finally realized that it is time to increase wealth in their communities and for their people and are open to the ideas of electronic commerce and the opportunities that are tied in to it (OECD, 2000).

It is essential, that developing countries increase their access to the Internet and to electronic commerce for their ability to perform in the constant growing global market. The expansion of electronic commerce to developing countries will automatically enable their overall economy.

The acceptance of electronic commerce all over the world is giving everyone the opportunities for advancement in all levels of society. The most important goal is giving developing countries the chance for better living conditions, increasing job opportunities, making services available, and providing better education and greater economy wealth.

# REFERENCES

African Development Forum (1999). *Electronic commerce in Africa.* 26, section 5.5.

Androuchko, L. (1998). Mozambique unveils leading edge telemedicine facility. *International Telecommunications Union*, Report ITU/98-1, 30 January 1998 Available online at: http://www.itu.int/newsarchive/press/releases/1998/98-01.html.

Ariyadasa, C. (2000). Sampath bank makes surprise move into stockbroking. *The Sunday Times*, Sri Lanka, 12 November 2000. Available online at: http://www.lacnet.org/suntimes/001112/busm.html.

Coeur de Roy, O. (1997). The African challenge: Internet, networking and connectivity activities in a developing environment. *Third World Quarterly.* *18*(5), 883-898.

Cordelia, L. (1999). What's holding back electronic commerce in Malaysia?. *New Straits Times- Management Times*. August.

Gagné, P. (1998). A special ITU development initiative. International Telecommunications Union Bureau for Telecommunications Development, article for *ITU News*, No. 3/98.

Goh, K. (1999). ITU teams up to push e-biz in developing world. *IDG News Service,* Boston Bureau, September 29.

Hay, P., Jackson, S., & Case McMahon, C. (2000). Developing countries could see fastest growth in over a decade but are hurt by trade barriers in rich

Copyright © 2003, Idea Group Inc. Copying or distributing in print or electronic forms without written permission of Idea Group Inc. is prohibited.

nations, Worldbank News Release No: 2001/126/S. Available online at: http://wbln0018.worldbank.org/news/pressrelease.nsf/673fa6c5a2d 50a67852565e2006.

Hillebrand, M. (1999). Asian electronic commerce set to go. *Electronic commerce Times,* May 6.

International Telecommunications Union (1999). Electronic commerce for developing countries. Report on the electronic commerce survey conducted in the framework of World Telecommunication Day 1999. Available online at: http://www.itu.int/newsarchive/wtd/1999/report.html.

National Law Center for Inter-American Free Trade (2000). Panama: proposed regulation of electronic commerce. Available online at: http://www.natlaw.com/ecommerce/docs/stpnec1.htm.

Ntoko, A. (1999). ITU to team up with world trade center and world internet secure key to expand electronic commerce in developing countries. *Telecommunication Development Bureau.* Report ITU/99-12, 17 September. Available online at: http://www.itu.int/newsarchive/press/releases/1999/12.html.

Ntoko, A. (2000). E-business infrastructure being deployed in more than 80 countries under ITU's electronic commerce for developing countries (EC-DC) project. *Telecommunication Development Bureau,* 14 July 2000. Available online at: http://www.itu.int/newsarchive/press/releases/2000/15.html.

Ochiengs, C. & Amey, C.F. (1998). Electronic commerce for developing countries (EC-DC): a special ITU development initiative. *International Telecommunications Union.* Available online at: www.itu.int.

Organization for Economic Co-operation and Development (OECD) (2000). The *Economic and Social Impacts of Electronic Commerce: Preliminary Findings and Research Agenda,* ISBN 9264169725. Available online at http://oecdpublications.gfi-nb.com/cgi-bin/OECDBookShop.storefront/EN/product/931999011P1.

Richman, D. (1999). Advocates hope for fair trade electronic commerce: developing countries' restrictions a concern. *Seattle Post-Intelligencer,* November 26. Available online at: http://seattlep-i.nwsource.com/business/ecom26.shtml.

Sadwosky, G. (1996). The internet society in developing countries. Nov/Dec 1996. Available online at: www.isoc.org/oti/articles/196/sadwosky.html.

Sivapalan, V. (2001). Need to rethink business. *New Straits Times- Management Times.* May 2001.

Copyright © 2003, Idea Group Inc. Copying or distributing in print or electronic forms without written permission of Idea Group Inc. is prohibited.

Sri Lanka Telecom (2001a). SLT sponsors National Institute of Education's pioneering effort in borderless education. 8 January, Available online at: http://www.slt.lk/inpages/newshiglights/26_sltsponsers.htm.

Sri Lanka Telecom (2001b). SLT Conducts A Seminar On Electronic commerce, 26th April 2001, http://www.slt.lk/inpages/newshiglights/39_sltconducts.htm

Turban, E., Lee, J., King, D. & Chung, H. M. (2000). *Electronic Commerce: A Managerial Perspective*. New Jersey: Prentice-Hall, Inc.

U. S. Department of Commerce - National Trade Data Bank. (1999). *Panama Marketing U. S. Products And Services*. Available online at: http://www.tradeport.org/ts/countries/panama/market.html.

Copyright © 2003, Idea Group Inc. Copying or distributing in print or electronic forms without written permission of Idea Group Inc. is prohibited.

**Chapter III**

# Adverse Effects of
# E-Commerce

Sushil K. Sharma
Ball State University, USA

Jatinder N.D. Gupta
University of Alabama in Huntsville, USA

## ABSTRACT

*E-commerce is the fastest growing industry worldwide and is one of the most rapidly evolving areas of national and international trade. The Internet has become an incredibly powerful tool for conducting business electronically. Companies have taken the proactive approach and are jumping on the new way to conduct business. E-commerce enables organizational change and helps organizations to conduct business with improved efficiencies and productivity. E-commerce is credited with empowering employees and knowledge workers, by giving them easy access to virtually unlimited information. E-commerce technologies have helped nations to accelerate their economic growth and to provide more opportunities for the businesses to grow. Meanwhile, it has also created many challenges and adverse effects, such as concerns over privacy, consumer protection, and security of credit card purchases, displacement of workers (especially low-status ones), and is charged with having a negative impact on quality of work life. This chapter describes the various adverse effects that have accompanied the advent of the Internet and e-commerce revolution.*

Copyright © 2003, Idea Group Inc. Copying or distributing in print or electronic forms without written permission of Idea Group Inc. is prohibited.

# INTRODUCTION

The Internet heralds an unprecedented evolution in the transformation of all business and communication. In 1991, the Internet had less than 3 million users around the world and its application to electronic commerce (e-commerce) was non-existent. By 1999, an estimated 250 million users accessed the Internet and made purchases online worth approximately $110 billion (Anonymous, 2000). E-commerce is defined as buying and selling of information, products, and services via computer networks or Internet (Anonymous, 2000, Sharma & Gupta, 2001). Electronic commerce promises to be the momentum behind a new wave of economic growth (Mariotti & Sgobbi, 2001). Internet and electronic commerce technologies are transforming the entire economy and changing business models, revenue streams, customer bases, and supply chains (Green, 2001). New business models are emerging in every industry of the New Economy. In these emerging models, intangible assets such as relationships, knowledge, people, brands, and systems are taking center stage (Boulton et al., 2000; McGarvey, 2001).

The relationship and interaction of customers, suppliers, strategic partners, agents, or distributors has been entirely changed. E-commerce has already changed the way traditional business transactions are conducted. It has improved business value by fundamentally changing the ways products are conceived, marketed, delivered, and supported. Companies are using the Internet as a medium to improve the quality of their customer relationships, whether by delivering better service through an e-mail bulletin board, or by lowering costs through the network enables just-in-time inventory control. According to the Forrester Research Group, in 1997, U.S. Internet commerce accounted for $8 billion goods and services. By 2002, Internet commerce is projected to rise to $327 billion (Penbera, 1999).

The Internet's growth and e-commerce has begun to create fundamental change to government, societies, and economies with social, economic and political implications. While the Internet revolution has created enormous upheaval for business by offering e-commerce solutions, it has also created plenty of opportunities for individuals and businesses in the new economy. On the positive side, e-commerce offers an opportunity for organizations to conduct business through a variety of new business models that help businesses to sell their products and services on line to consumers. E-commerce is helping organizations to reduce transaction, sales, marketing, and advertising costs. E-commerce is also helping businesses to reach global markets at low costs to conduct business 24 hours per day, 7 days per week, 365 days per year. The value of e-commerce was estimated at around $650 billion worldwide in 2000 (Bassols & Vickery, 2001). According to Frank Gens, senior vice president of Internet Research at Analyst Company IDC, the e-commerce market will be worth $900 billion by the year 2003. Gens

Copyright © 2003, Idea Group Inc. Copying or distributing in print or electronic forms without written permission of Idea Group Inc. is prohibited.

predicts that customers will spend about $2 million a minute on the Internet (Town, 1999).

The impact of e-commerce is realized in many different ways both positive and adverse. Many of the benefits come from improved consumer convenience, expanded choices and lower prices. E-commerce has provided opportunity for better interactions with partners, suppliers, and targeted customers for service and relationships. E-commerce can improve product promotion through mass-customization and one-to-one marketing. It offers a new direct channel for selling existing products, reduces the cost of some processes (e.g., information distribution), and reduces the time to market. E-commerce also improves customer service through automated services and round-the-clock operation, providing the customers with choice, information, convenience, time, and savings with improvements that add value to their shopping. E-commerce can also be used as an interactive medium that fosters social interaction. Many consumers and sellers use the Internet not only as a tool for information gathering and e-commerce, but also view the Web as a means to keep in touch and interact with each other. People often prefer web surfing to watching TV because it is a more interactive medium for entertainment on the Internet.

Although e-commerce has provided a number of opportunities and benefits to customers and businesses, concerns have surfaced about privacy, security, frauds, consumer protection, and abuse of personal information. Many equate the loss of privacy with loss of personal freedom (Borck, 2001). On the other hand, e-commerce also affects such areas as composition of trade, labor markets, taxation, and prices. Some of these effects of e-commerce are unintentional and create adverse business and personal conditions that could have societal consequences. This chapter describes the various adverse effects or problems created by the advent of the Internet and e-commerce revolution. While the adverse effects of e-commerce and the Internet do not fit into mutually exclusive categories, we categorize them into three hierarchical levels: societal and economic, business, and individual and focus discussion on the adverse affects each of these levels.

# ADVERSE SOCIETAL AND ECONOMIC EFFECTS

As e-commerce continues to grow rapidly, it could have significant effects on the structure and functioning of a society at an aggregate level. The impacts of these changes are diverse and likely to widen the digital divide among nations, alter the composition of trade, disrupt labor markets, and change taxation (Anonymous,

Copyright © 2003, Idea Group Inc. Copying or distributing in print or electronic forms without written permission of Idea Group Inc. is prohibited.

2000). In this section, we briefly discuss some major adverse societal impacts of e-commerce.

## Widening Digital Divide

The term *Digital divide* means a lack of equal access to computer technologies and the Internet in particular, creating a gap between those who *have* and those who *have not*. One dimension of the digital gap follows demographics of gender, race, and social class. Another dimension follows the economic gap between wealthy and poor countries. Millions of technologically disenfranchised have-nots, who cannot afford the cost of that technology and training, are walled off from potentially life-changing tools and knowledge. Therefore, they feel isolated in the virtual world. For example, although growth has been very strong in Europe, particularly in Sweden and Finland, the United States still accounts for more than three-quarters of all e-commerce transactions. Despite the promise of "borderless" trade, most e-commerce is still national or within the continents (Bassols & Vickery, 2001).

The digital divide gap between the technology *haves* and *have-nots* presumably is also across white, wealthy, and urban Americans with computers and Internet access on the one hand and minority, poor, and rural Americans who lack computers and Web access on the other (Quay, 2001). Today the real digital divide is the chasm between those who use Internet technologies to improve their lives and those who do not. Another digital divide is based on geography. Disparities in the location and quality of Internet infrastructure, even the quality of phone lines, have created gaps in access (Quay, 2001). Hindered by poverty and a poor telecommunications infrastructure, Internet penetration rates in the developing regions range from less than 1 percent to 3 percent, which is far below the 25 percent to 50 percent penetration rates seen in many wealthy and developed nations (Rombel, 2000). E-commerce and access to the Internet create an enormous disparity in wealth and a great need to spread and disperse this wealth and connectedness for a socially sustainable benefit.

## A Threat of Information Warfare

In information warfare, hackers, by controlling an organization's computer systems, obtain a significant advantage by changing the flow of information, altering information, or shutting down a system completely. A computer system may not only be shut down, but also could be destroyed by infiltration and distribution of various computer viruses. The breakdown of major information systems such as banking networks or electricity grids could be a devastating event, altering the economy of a nation and the daily life of its population. While the online

Copyright © 2003, Idea Group Inc. Copying or distributing in print or electronic forms without written permission of Idea Group Inc. is prohibited.

environments such as e-commerce are good for offering online products and services, these environments pose a threat to governments. Online environments are vulnerable to attacks from inside and outside the companies. Many of the recent virus attacks triggered by hackers targeting specific sites indicate how hackers can affect businesses. A country where almost all the systems such as electricity, water, transportation, and businesses are automated becomes a major target for information warfare.

Many experts fear that abuse of online environments may become a weapon to overthrow governments. Information warfare is definitely a concern of most nations as can be seen by the large number of nations researching and developing offensive information warfare systems. This increased awareness of the potential damage cyber attacks can inflict has caused most nations to research defensive information warfare tactics (Sharma & Gupta, 2001).

## Abuse of Power

Computer matching is a technique employed to verify personal information by cross-referencing it with a separate, unrelated database. The dangers of computer matching include the abuse of power, the existence of outdated data and the possibility of information leaks. In 1988, the United States Congress passed the Computer Matching Privacy and Protection Act. The legislation permitted government agencies to compare information from two or more sources to detect anomalies. If information stored in different databases varies, people may be denied certain privileges based on wrong records or outdated data. Some fear that the results of computer matching are as effective as door-to-door searches and can indict them for mistakes of others if wrong data is used. It again leads to invasion of privacy (Shattuck, 1996).

Customers may give false personal information while shopping or surfing on the net to avoid being profiled by marketing companies. Due to privacy concerns, people may intentionally provide incorrect information about themselves, creating problems by having wrong information recorded in some databases. Incorrect information can create misery when used and exchanged by various agencies with banks, and health organizations. Incorrect information can cost a job, opportunity or the denial of a bank loan or mortgage. Data can be misconstrued in a variety of strange ways. If information is not verified properly, it can lead to drastic results. The verification of information by cross-referencing can also lead to positive consequences for both individuals and organizations. Therefore, its negative impact must be minimized so that individuals and organizations can benefit from its positive results.

Copyright © 2003, Idea Group Inc. Copying or distributing in print or electronic forms without written permission of Idea Group Inc. is prohibited.

## Impact on Children

The interactive and multimedia nature of e-commerce technologies provides tremendous opportunities for children to access resources to attain educational goals. At the same time, it presents unique challenges for protecting the privacy of young children. Since the Internet allows children to buy items online, it exposes them (children) to messages that are more commercial. Children may be required to fill out questionnaires in order to enter sites, join clubs, play games, move to "special" portions of a Web site. They may be invited to give personal information in the course of participating in a chat room. During their online buying, children may innocently provide information that can lead to receiving undesired material (pornography) or junk mail. This adds another burden for parents to monitor their children' surfing on the net.

## Impact on Economic Performance

A number of studies have attempted to quantify the impact of e-commerce at the macroeconomic level. Some studies credit e-commerce for an increase in the level of GDP by 2 to 5 percent (Penbara, 1999, Anonymous, 2000). These studies, however, are based on a number of quite restrictive assumptions and their results cannot be interpreted as a positive indication. In assessing the implications for macroeconomic policy, it should be borne in mind that the Internet also boosts aggregate demand.

Friedman (1999) has argued that Internet related technologies could increase the speed of financial operations, which raises the issue as to how interest rates should be set and whether the short end of interest setting needs to become shorter, i.e., time units smaller than a day. Some economists have even envisaged a world where technological developments emasculate altogether the monetary controls of central banks (King, 1999). This could occur if new technologies (and regulators) permitted real time pricing and exchange of goods across the Internet without the intercession of an independent monetary system administered by a central bank. In such an environment, the government earns no seignorage and would no longer be able to provide liquidity support by printing money.

In the event that electronic monies do gain a sizeable share of payment systems, their close substitutability with other payment instruments raises issues about the definition of monetary aggregates, their stability and the ability for central banks to control money supply. Moreover, seigniorage revenues accruing to central banks could fall. Another concern with electronic monies is the possibility that they will be used for money laundering (Anonymous, 2000).

Copyright © 2003, Idea Group Inc. Copying or distributing in print or electronic forms without written permission of Idea Group Inc. is prohibited.

## Encouragement of Monopoly Practices

Despite signs that monopolistic practices are increasing, including the more obvious vertical integration of firms within the information industry as a whole, there has been strong countervailing pressure against government intervention in Internet activities. Others believe that a government-inspired monopoly in e-commerce would be less onerous than a natural monopoly under the assumption that government-protected monopolies are easier to unwind. Of course, experience demonstrates that although monopolies aided by government are less obvious to the public, they still promote inefficiencies and are very difficult to eliminate (Penbera, 1999).

## Impact on Tax, Trade and Regulatory Policies

E-commerce has a strong impact on taxation and tax policy. Concerns have been expressed that e-commerce could result in the erosion of tax bases. Consumption taxes are levied on the principle of taxation at the place of consumption and according to rates set in individual countries, or in individual states in the case of federal nations. E-commerce, however, has the potential to undermine the application of domestic and national tax rules. Tax planning for an e-business differs from tax planning for a traditional bricks-and-mortar company. Historically, the generation of income depended on the physical presence of assets and activities. This physical presence, or permanent establishment, generally determined which jurisdiction had the primary right to tax the income generated. Because of the growth of electronic commerce, new e-business models (including digital marketplaces, on-line catalogs, virtual communities, subscription-based information services, on-line auctions, and portals) have emerged. Each allows taxpayers to conduct business and generate income in a country with little or no physical presence in that country. The separation of assets and activities from the source of the income represents a significant departure from historic business models. This change creates new tax planning challenges and opportunities (Olin, 2001; Anonymous, 2000).

A Few European countries suggest implementation of a tax called "bit tax" (i.e., a tax on the "bits" of information zooming around computer networks). These countries support such a tax, partly because Europe (with high rates of VAT) stands to lose the most from untaxed electronic sales. In America, which does not have a federal sales tax, the idea has not found much favor, and the present US administration rejects the idea of any new taxes on the Internet.

The basic problem with a 'bit tax' is that it is indiscriminate: it taxes not just on-line transactions, but all digital communications. In addition, the question of

Copyright © 2003, Idea Group Inc. Copying or distributing in print or electronic forms without written permission of Idea Group Inc. is prohibited.

valuation would be difficult to determine. More importantly, it is argued that duties will crush the development of e-commerce and stunt its growth. If implemented in some countries, it would simply drive business offshore and on-line transactions would simply take place in a state or country where there is no such tax. Since trade policy - like tax policy - is based on such distinctions, governments may find it difficult to determine jurisdiction and tariff revenue rights. Moreover, the laws and regulations that a consumer relies on for protection at home may not apply in the merchant's country. Indeed, in some quarters, there are concerns that the scope for the Internet to transcend national boundaries could emasculate the ability of regulatory bodies to fulfil their objectives. Thus, there is a need to update regulatory frameworks and strengthen co-operation between regulatory bodies to achieve the goals of economic regulations without jeopardizing the efficiencies likely to be associated with the growth of e-commerce (Penbera, 1999; Anonymous, 2000).

## Impact on Employment and Labor Policy

The growth of e-commerce is likely to have both direct and indirect impacts on labor markets as well as the composition of employment. Since e-commerce may create more knowledge-based products, it is likely to drive widespread changes in the labor market, shifting the composition of workers required to produce and deliver a product or service (Anonymous, 2000). There would be shift in kind of skills needed. Faster rates of innovation and diffusion may also be associated with a higher turnover of jobs. This may create more turbulence, as workers would need a skill upgrade from time to time. This may result into change in swift labor policies for reallocation of labor to the changing needs of the economy (Anonymous, 2000).

# ADVERSE BUSINESS EFFECTS

While e-commerce has enhanced business effectiveness and efficiency, it has created some adverse effects as well. This section discusses major adverse business effects of e-commerce.

## Emerging Monopolistic Trends

Since e-commerce transcends geographical boundaries, many big firms with known brands may not only expand their markets, but may also enter into new business activities across the broad spectrum of business activities. This may help to reduce the costs and prices, but it would have a greater danger of creating a monopoly of e-commerce by a few corporations or networks of corporations.

Copyright © 2003, Idea Group Inc. Copying or distributing in print or electronic forms without written permission of Idea Group Inc. is prohibited.

Many firms may use a low price strategy to grab the market and eliminate the competition. Several mergers and alliances, in which two or more firms combine to achieve a large market share and have large economies of scale, can result in eliminating meaningful competition (Penbera, 1999).

## Impact on Competition Policy

E-commerce offers the ability to expand markets. By expanding the size of the market, the e-commerce could create opportunities to dominate markets. Persistence of price dispersion across Internet markets and the absence of noticeable price reductions has led to concerns that the cost structure of some Internet markets, which could ultimately result in a less competitive industry (Anonymous, 2000).

The scope for noncompetitive behavior is perhaps strongest among "digital" and knowledge intensive products. For such products, once the first copy is produced (like a software application), the cost of a second copy is close to zero. Due to the large marketing capital needed to develop visibility and a brand name, start-up companies are finding it difficult to enter the market. Low contestability is resulting in highly concentrated scenarios where one big brand takes it all and this has made an impact on competition policy (Penbera, 1999; Anonymous, 2000).

## Impact on Labor Costs and Employment

E-commerce is facilitating the shift from the mass labor paradigm to a knowledge-worker paradigm. Productivity enhancements and labor cost savings are major driving forces behind e-commerce activities. E-commerce efficiencies have displaced thousands of clerical personnel in manufacturing and service sectors and created large gains in unemployment among less-skilled workers. This shift has enormous economic implications for more populated countries that have relatively cheap mass labor available and use cheap labor as strategy to improve their competitiveness and overall growth. The shift from mass to knowledge labor has already created a shortage of knowledge workers in several countries where the education system and technology infrastructure have not been very strong (Penbera, 1999).

## Impact on Prices

Electronic commerce is widely expected to improve efficiency due to reduced transaction and search costs, increased competition and more streamlined business processes. Lower search costs may also lead to Internet consumers being more sensitive to price changes. By reducing search costs and increasing the flow of information, e-commerce might effectively shift power from producers to consum-

Copyright © 2003, Idea Group Inc. Copying or distributing in print or electronic forms without written permission of Idea Group Inc. is prohibited.

ers and make it harder for firms to maintain higher prices (Bakos, 1997). However, empirical evidence does not support this claim in all cases. Brynjolfsson and Smith (1999) found that average prices on certain items in a particular industry sold through the Internet were lower than their equivalent purchased through traditional retailers. However, in certain cases prices of goods sold through the Internet were higher than those charged by traditional retailers. Brynjolfsson and Smith (1999) justified this phenomenon by arguing that certain reductions in cost are offset by higher overhead costs elsewhere. They also indicate that increases or decreases in price depend on the size of the market.

## Cyber Slacking

Businesses worldwide are clamoring to get their hands into the 'e-world' and all its benefits. Most companies today view the Internet as an essential tool for their employees. The Internet is not just regarded as a method of keeping up-to-date, but as a method of disseminating company wide information and flattening the corporate structure. Many companies are also turning to e-mail and networked conferences as a method of keeping their global divisions coordinated and current. These technologies are implemented in the hope of seeing productivity rise and quality increase but companies have not seen corresponding productivity increases. Employees use the Internet for entertainment and other personal uses. This activity has been coined as *cyber slacking*. In the new-wired world, cyber slacking is an important social and economic issue. Its effects are seen in debates about computer productivity, Internet censorship, computer monitoring, legal considerations and managerial challenges.

Companies equip their employees with access to the Internet to increase productivity. Therefore, cyber slacking is especially troublesome to these companies as the tools that were bought to increase productivity can often be found to lessen it. Companies must often police Internet use, which has resulted in some publicized firings of involved employees. This has increasing societal implications on companies. As companies crack down on inappropriate use of the Internet, more people feel threatened at work. Thus, an environment of tension, paranoia and "shoulder-looking" easily arises. This affects the way employees feel *at* work and about *going* to work (Gupta & Sharma, 2002).

## Direct Mail Marketing

Most people have experienced the menace of direct mail when they receive undesired mail from unknown persons or companies. When people make online purchases, companies may ask them to disclose personal information and preferences through surveys that are available online. Business web sites ask visitors to

Copyright © 2003, Idea Group Inc. Copying or distributing in print or electronic forms without written permission of Idea Group Inc. is prohibited.

supply personal data, particularly when information, services or merchandise are requested. Companies store information on consumers and profile them in order to market items that may be of interest to them.

Some organizations sell collected information as a good or service to other potential marketers (Ackerman et al, 1996). Companies may also mine data from public records like phone books, automobile registrations, driver's licenses or home ownership documents located in city halls. This information is garnered by companies and in a database for marketing purposes (Hatch, 1996). This practice can lead to the endless junk mail that was never requested. Direct mail marketing is a form of privacy invasion. Junk e-mail is becoming very pervasive, with one bulk e-mailer, Cyber Promotions, boasting that it sends 1.5 million messages a day (Zaret & Sawyer, 2000).

# ADVERSE INDIVIDUAL EFFECTS

E-commerce can significantly affect individual freedom and living conditions. The adverse effects of e-commerce and Internet on the individual include social isolation, loss of security and privacy and loss of individuality. In this section, we briefly discuss these adverse individual impacts of e-commerce.

## Social Isolation

E-commerce has far reaching implications in a social context. On one hand, it provides all the comfort of shopping from home; on the other hand, it removes old-fashioned human interactions for social needs. Take for instance the case of telecommuting that is becoming a reality. Today many organizations allow their employees to work from their homes. It is predicted that by 2010, the separation between work and home will have blurred, and areas once zoned for commercial use will be converted into live-work condominiums. The primary motivator for worker participation in telework programs is the desire to increase overall productivity through autonomy. Employees are able to choose their work hours in which they feel they are most efficient and can reduce the number of interruptions by co-workers and the stress of daily commuting. Therefore, they have an increased desire to work. In addition, individuals generally have more free time and are able to work overtime, without spending long hours away from their families.

There are some positive aspects of telecommuting to organizations and society. With increased employee autonomy, organizations are able to retain their employees by providing more flexibility in their work schedule. Organizations benefit from fewer costs in recruiting, training, and disability costs. Organizations are also able to locate satellite offices in less densely populated areas. Environmen-

Copyright © 2003, Idea Group Inc. Copying or distributing in print or electronic forms without written permission of Idea Group Inc. is prohibited.

tal issues decrease with the elimination of carbon dioxide from commuting cars. Quality of life in general increases where parents are able to spend more time with their children and can play a more dominant role in their community. Nevertheless, the negative aspect of telecommuting should not be ignored. The chief problem appears to be the fear of losing touch.

The stimulation of interacting with colleagues may be lost and the resultant gradual, social isolation may affect opportunities for promotion and the selection of career-advancing assignments. Improved conditions and working from home will be limited to those wearing white collars. The working classes will probably remain 9-to-5ers in hands-on, site-specific jobs. Therefore telecommuting can increase the gap between social classes. Telecommuting could be seen as a different social class with a different set of rules. The fact that traditional workers could be seen as failures in such a society further increases rifts between social classes.

As the country's young people age, and gain employment, they will increasingly use e-commerce to purchase many consumer goods. This will serve to limit the need for new regional shopping malls and strip-commercial centers. There will be increased pressure for warehousing and distribution centers to meet the new demand created by e-commerce (Kemp, 2001).

## Loss of Security

E-commerce offerings not only provide new opportunities for customers and businesses but also open companies up to security vulnerabilities. Privacy and security concerns in e-commerce are under intense pressure from consumers, lawmakers and regulators to provide foolproof security safeguards and policies to protect their systems and customer privacy (Miyazaki & Fernandez, 2000).

Security refers to the integrity of the data storage, processing and transmitting systems and includes concerns about the reliability of hardware and software, the protection against intrusion, or infiltration by unauthorized users. While buying online, people naively believe that their communication is private and secure. Unfortunately, in many cases, that is not true and messages are insecure and vulnerable to hackers. Any message sent on the Internet typically travels through many computers. At each computer, it may be intercepted, read and even changed. Skilled hackers may tap into communication particularly if it is financial in nature and can benefit them. In spite of a number of security technologies used to protect communication, expert hackers can still access sensitive data.

A variety of security breaches commonly take place in today's world and often are not reported by organizations. In the 2001 Computer Crime and Security Survey (Computer Security Institute/Federal Bureau of Investigation, 2001), 91 percent of respondents said they had experienced a security breach, but only 36

Copyright © 2003, Idea Group Inc. Copying or distributing in print or electronic forms without written permission of Idea Group Inc. is prohibited.

percent reported the incident to law enforcement. The 1997 Computer Security Institute/Federal Bureau of Investigation survey on computer crime and security revealed that 75 percent of respondents had reported financial losses due to various computer security breaches. These breaches ranged from financial fraud, theft of proprietary information, sabotage and computer viruses on the high end to Internet abuse and laptop theft on the low end (Rapalus, 1997; Morgan, 2001). Reports indicate that hackers have penetrated the security systems of many companies including America Online, the world's biggest online service provider. Forrester Research estimates that $12.2 billion in e-commerce revenue was lost to privacy concerns in 2000 after $2.8 billion was lost in 1999 (Rombel, 2001).

Several bankrupt online retailers in Europe and other companies were reportedly selling customer lists for much needed cash despite earlier promises in privacy statements never to do so (Rombel, 2001). Many network vulnerability scanners and intrusion detection systems have been developed and implemented but systems still seem to be vulnerable.

## Loss of Privacy

Privacy is defined as an individual's right to be left alone, free from interference or surveillance from other parties. On one hand, the Internet has provided opportunity for conducting business online in the form of e-commerce, but on the other hand, organizations may be secretly profiling and collecting information about customers. Spamming, which is the practice of sending out unsolicited e-mail, is growing because it costs so little to send out millions of messages or advertisements electronically. Many prominent high-technology companies have already been caught attempting to quietly collect information about their customers via the Internet (Gupta & Sharma, 2002).

Privacy has become a key issue in the digital age. Technological advances make it easy for companies to obtain personal information and to monitor online activities, thus creating a major potential for abuse. There are three areas of concern in the privacy debate: employers monitoring employee computers and Internet use in the workplace, advertising and market research companies collecting and selling personally identifiable information based on consumers' online activities, and information brokers selling readily available personal information from public-record databases online (Blotzer, 2000).

Most companies collect information about the visitors who visit a company's site. Files such as "**cookies**" are planted on a computer by the web sites that are surfed so that they track down the surfing details of users. Most users are blissfully unaware of the electronic footprint they leave when they surf web sites. Cookies are short pieces of data used by web servers to help identify web users. Cookies are

Copyright © 2003, Idea Group Inc. Copying or distributing in print or electronic forms without written permission of Idea Group Inc. is prohibited.

used to identify a web user and to track their browsing habits. Cookies help companies to prepare marketing databases that enable companies to better sell their products and services to consumer demands. By performing statistical analysis of database information on consumers, companies prepare direct mailer lists. Companies collect information about customers' buying habits and preferences so they can customize their products and services. At times, customers blame companies for infringing on their privacy by collecting the information through cookies. While cookies may be desirable from a company's point of view, customers feel that organizations collect too much private information and may attempt to sell it to a third party of potential marketers. The ability of marketers to track surfing and buying patterns can and does lead to abuses.

## E-Mail Monitoring

The use of e-mail in the workplace produces many conflicting opinions in terms of its use and abuse both by the user (i.e., the employee), and the service provider (i.e., the employer). This causes confusion with the legal and ethical guidelines that must be followed with this communication medium in order to maintain both the rights of the user (i.e., privacy) and the rights of the employer (i.e., monitoring). The primary goal of a corporation in monitoring its employee's e-mails is to prevent those that are either vulgar, offensive, or those that compromise the company's best interests (i.e., breach of confidentiality). However, providing managers with the means to monitor their employees can result in an abuse of power. It creates confusion among employees and corporations as to where their loyalties lie (Ambrose & Gelb, 2001).

## Loss of Individuality

Maintaining a customer base has become very important asset in today's economy for the organizations to gain competitive advantage. Therefore, organizations use sophisticated tools to reach customers and get their personal data recorded into their databases. This helps organizations to customize the products and services as per the needs of the customers. In the coming years, organizations will continue to seek personal information about their employees and customers since it is a source of power. Many organizations use cookies or other technological tools to collect information about their customers, their preferences and items they wish to buy. Organizations use data-intensive techniques like profiling and data mining to aid in precision marketing or to customize products for one-to-one marketing. This information could also aid in connecting people of similar interests effectively forming various new world wide social clubs and hobby groups.

Copyright © 2003, Idea Group Inc. Copying or distributing in print or electronic forms without written permission of Idea Group Inc. is prohibited.

Companies can sell their customer information as a good or service much as if a magazine publisher would sell its mailing lists.

Many believe that e-commerce technology is eroding personal privacy because consumers have no control over their personal data that merchants have collected during their shopping experience. In addition, personal record-keeping systems of merchants are not regulated or restricted. People fear that if the trend of collecting information continues, they may lose their individuality since they would have no control over the information about them (Kling & Linowes, 1996).

# CONCLUSIONS

E-commerce involves buying and selling on the Internet among businesses and consumers. E-commerce has provided opportunity for better interactions with partners, suppliers, and targeted customers for service and relationships. E-commerce provides the customers with choice, information, convenience, time, and savings with improvements that add value to their shopping. However, loss of privacy with e-commerce, security issues, increasingly sophisticated frauds, abuse of personal information, and impact on prices have aroused as the main concerns.

Companies increasingly rely on the collection and use of information about consumers for many purposes, including targeted advertising and marketing, maximizing the convenience of electronic commerce, and personalizing customer service and support. In many ways, the creation and maintenance of databases containing information about consumers and their activities can provide a "win-win" situation for both private companies and individuals. It enables companies to operate more efficiently, lower costs, and streamline marketing and promotion efforts while simultaneously giving consumers access to "customized" information about products and services that correlates to their personal preferences. The collection and use of personally identifiable information has, however, raised significant concerns from lawmakers, regulators, and private litigants, especially when it involves sensitive information such as financial or medical information and information related to children.

The e-commerce revolution, affordability, technology and the skill to use has been restricted to the world's information haves. E-commerce infrastructure is still not available in many countries and even if available in some countries, it is limited to a few cities. Much of the world's population is part of the information have-nots —those who lack the technology and training of the technologically privileged. E-commerce promoters must also be challenged to try to bridge the gap between these "have and have-nots" in order to form a singular society and not a society separated by a digital divide.

Copyright © 2003, Idea Group Inc. Copying or distributing in print or electronic forms without written permission of Idea Group Inc. is prohibited.

The e-commerce and Internet revolution could adversely affect the ability of a society, business or an individual in various ways. As e-commerce and Internet use becomes a common household and business phenomenon, strategies for minimizing the adverse effects of such impacts must be developed and implemented. It is only then that we can realize the full potential of these emerging powerful technologies and philosophies.

# REFERENCES

Ackerman, M. S, Allen, J. P. & Kling, R. (Eds.) (1996). *Information Entrepreneurialism and Controversy* (2nd ed.), San Diego: Academic Press.

Ambrose, Jr., S.F. & Gelb, J.W. (2001). Consumer privacy regulation and litigation. *The Business Lawyer, 56* (3), pp. 1157-1178, Chicago.

Anonymous (2000). E-commerce: Impacts and policy challenges. Organization for Economic Cooperation and Development. *OECD Economic Outlook,* (67), 193-213.

Bakos, J.Y. (1997). Reducing buyer search costs: Implications for electronic marketplaces. *Management Science, 43,* (12), pp. 1676-1692.

Bassols, V.L. & Vickery, G. (2001). E-commerce: The truth behind the web. *Organisation for Economic Cooperation and Development. The OECD Observer,* 24, pp. 17-19.

Borck, J.R. (2001). Currency conversion, fraud prevention are hurdles to successful global commerce, *InfoWorld, 23* (6), p. 55.

Boulton, R.E.S., Libert, B.D. & Samek, S.M. (2000). A business model for the new economy. *The Journal of Business Strategy , 21* (4), pp. 29-35.

Computer Security Institute/Federal Bureau of Investigation (2001). Computer crime and security survey. *Computer Security Issues & Trends, 7* (1).

Green, P.L. (2001). E-finance business models evolve, *Global Finance, 15* (3), pp. 30-35, New York.

Gupta, J.N.D & Sharma, S.K. (2001). Cyber shopping and privacy. In A. Gangopadhyay (Ed.), *Managing Business with Electronic Commerce: Issues and Trends,* Hershey, PA: Idea Group Publishing, pp. 235-249.

Hatch, D. (1996). Privacy: How much data do direct marketers really need. In R. Kling (Ed.), *Computerization and Controversy* (2nd ed.), San Diego: Academic Press.

Kemp, R.L. (2001). Cities in the 21st century: The forces of change. *Economic Development Review, 17* (3), pp. 56-62.

Copyright © 2003, Idea Group Inc. Copying or distributing in print or electronic forms without written permission of Idea Group Inc. is prohibited.

Kling, R. (1996). Information technologies and the shifting balance between privacy and social control. In R. Kling (Ed.), *Computerization and Controversy: Value Conflicts and Social Choices*, (2nd ed.), San Diego, CA: Academic Press, pp. 614-636.

Linowes, D. F. (1996). Your personal information has gone public. In R. Kling (Ed.), *Computerization and Controversy: Value Conflicts and Social Choices*, (2nd ed.), San Diego, CA: Academic Press, pp. 637-642.

Mariotti, S. & Sgobbi, F. (2001). Alternative paths for the growth of e-commerce. *Futures, 33* (2), pp. 109-125, Kidlington.

McGarvey, R. (2001). New corporate ethics for the new economy, *World Trade, 14* (3), p. 43.

Miyazaki, A.D. & Fernandez, A. (2000). Internet privacy and security: An examination of online retailer disclosures. *Journal of Public Policy & Marketing, 19* (1), pp. 54-61.

Morgan, L. (2001). Be afraid...be very afraid—Malicious attacks are on the rise, and trends are harder to predict-step one is admitting your company is vulnerable. *Internetweek*, 843, pp. 37-38.

Penbera, J.J. (1999). E-commerce: Economics and regulation. *S.A.M. Advanced Management Journal, 64* (4),  pp. 39-47.

Olin, J. (2001). Reducing international e-commerce taxes. *World Trade, 14* (3), 64-67, Troy.

Quay, R. (2001). Bridging the digital divide. *Planning, 67* (7), pp. 12-17, Chicago.

Rapalus, P. (1997). Security measures for protecting confidential information on the Internet and intranets. *Employment Relations Today, 24*, pp. 49-58.

Rombel, A. (2000). The global digital divide. *Global Finance, 14* (12), p. 47.

Sharma, S.K. & Gupta, J.N.D (2001). E-commerce opportunities and challenges. In M. Singh and T. Teo (Eds.), *E-Commerce Diffusion: Strategies and Challenges,* Australia: Heidelberg Press, pp. 21-42.

Shattuck, J. (1996). Computer matching is a serious threat to individual rights. In R. Kling (Ed.), *Computerization and Controversy* (2nd ed.), San Deigo, CA: Academic Press.

Town, B. (1999). E-commerce - tomorrow's marketplace, *Online & CD-ROM Review, 23* (2), pp. 107-109.

Zaret,-E. & Sawyer, S. (2000). Protect yourself online. *Macworld, 17* (7), pp. 64-69.

Copyright © 2003, Idea Group Inc. Copying or distributing in print or electronic forms without written permission of Idea Group Inc. is prohibited.

**Chapter IV**

# The Emerging Need for E-Commerce Accepted Practice (ECAP)

G. Erwin
Cape Technikon, South Africa

S. Singh
University of South Africa, South Africa

## ABSTRACT

*With the rapid expansion of the Internet and constant technological advances, it is clear that e-commerce will reshape methods of the business world. Government, large corporations, medium and small business now have to conduct their electronic activities in an accountable, transparent and well-structured way. Unlike traditional business recording, such as accounting, with Generally Accepted Accounting Practice (GAAP), no guidelines or frameworks exist that recognize e-commerce issues.*

Copyright © 2003, Idea Group Inc. Copying or distributing in print or electronic forms without written permission of Idea Group Inc. is prohibited.

# INTRODUCTION

*"The newest innovations, which we label information technologies, have begun to alter the manner in which we do business and create value, often in ways not readily foreseeable even five years ago."*

*Alan Greenspan*
*Chair, Federal Reserve Board*
*May 6, 1999*

A large multinational corporation once attempted to sell baby food in an African nation by using packaging designed for its home country market. The company's regular label showed a picture of a baby with a caption describing the kind of baby food contained in the jar. African consumers took one look at the product and were horrified. They interpreted the label to mean that the jar contained ground-up babies (Ricks, 1993)!

This example serves to illustrate that the design of effective e-commerce sites requires careful planning and sensitivity to cultural issues. For example, South Africa is a low to middle-income, developing country with an abundant supply of resources, well-developed financial, legal, communications, energy and transport sectors, a stock exchange that ranks among the ten largest in the world, and a modern infrastructure supporting an efficient distribution of goods to major centers throughout the region. However, growth has not been strong enough to reduce the prevailing 30 percent unemployment figure. Daunting economic problems remain from the apartheid era, especially poverty and lack of educational and economic empowerment among disadvantaged groups. At the end of 2000, President Mbeki vowed to promote economic growth and foreign investment, and to reduce poverty by relaxing restrictive labour laws, stepping up the pace of privatisation, and cutting unneeded governmental spending. Eleven official languages and cultural diversity further complicate all these challenges. E-commerce cannot target specific audiences, so South Africa illustrates how difficult it is to appeal to customers.

The evolution of Web-based business activity has resulted in the term e-Business referring to three categories of business activity:

- Business-to-Employee (B2E): Intranet-based applications internal to a company.
- Business-to-Consumer (B2C): Internet-based applications for a company's customers and
- Business-to-Business (B2B): Extranet-based applications for a company's business partners. (An IOS: Inter Organisational System).

Copyright © 2003, Idea Group Inc. Copying or distributing in print or electronic forms without written permission of Idea Group Inc. is prohibited.

[The term *e-Commerce* often refers only to B2B (McNurlin & Sprague, 2002)]

# BACKGROUND

The use of information technology (IT) - computers and telecommunications - has been growing at an increasing rate ever since the invention of the modern computer in the 1950s. Now, with the widespread development of the Internet and the World Wide Web (WWW, 'the Web'), change has accelerated even more (McNurlin & Sprague, 2002). Computer-based information systems (CBIS) are information systems that require hardware, software, databases, telecommunications, procedures and people to accomplish their goals (Stair, 1992).

As a preliminary investigation into good Web practice, the authors conducted research on technological features that affect Web-based services (Erwin & Singh, 2001).

Some comments by respondents to a survey on Web-based service features were:
- 'I consider it important that in addition to Web sites with all the "bells and whistles", a "text-only" version should be made available for fast load times and for learners who do not have fast links and the latest browsers.'
- 'Tools need to support doing the basic functions quickly and easily rather than doing lots of fancy stuff.'
- 'I (and most colleagues) hate frames, therefore tend to avoid them.'
- 'Often "help" isn't helpful!'

These quotations indicate resistance to new technologies (methods) and propensity to use a relatively small subset of available features.

Tools supporting e-commerce often have older technologies such as telephones, faxes and help desks. However, there is an emerging mix of technologies specific to the Web (Erwin & Singh, 2001). These technologies now introduce a new set of complexities for an auditor of this electronic environment.

# TECHNOLOGICAL TRENDS IN INFORMATION TECHNOLOGY

Technological changes are having enormous impact on the capabilities of organizational systems (Turban, et al., 2001). Table 1 outlines some general information technology trends affecting organizational systems.

Copyright © 2003, Idea Group Inc. Copying or distributing in print or electronic forms without written permission of Idea Group Inc. is prohibited.

*Table 1: Source: Turban et al. (2001)*

| Trend | What It Is | Benefits |
|---|---|---|
| Data warehousing | Gigantic computer 'warehouses' (storage) of large amounts of data. | Data warehouses organize data for easy access by end users of the data. When integrated with the Internet, they can be accessed from any location at any time. |
| Data mining | A sophisticated analysis technique that automatically discovers previously undetected relationships among data. | Enables managers to see relationships and dynamics in data elements that they had not foreseen (e.g., how the sales of one product might drive the sales of another product). |
| Intelligent systems and agents | Automated rules that execute preprogrammed decisions or tasks when encountering specified conditions in data. | Increase productivity and ease the execution of complex tasks. Intelligent agents help users navigate the Internet, access databases and conduct electronic commerce. |
| Electronic commerce | Business done online; the exchange of products, services and money with the support of computers and computer networks. | Can provide a competitive edge and could change organizational structure, processes, procedures, culture and management. |
| Electronic document management | A technique that converts paper-based documents to digital electronic form via scanning and related technologies. | Greatly reduces storage requirements and allows the documents to be organized and manipulated like any other type of electronic data. |

Every item shown in Table 1 is a potential source of attributes of a legacy system. E-commerce is a composite of the trends in Table 1.

# THE WORLD WIDE WEB (WWW) – THE WEB

The WWW as a vehicle for the implementation of trade and commerce has attracted the attention of business and government. There are many applications of the WWW, such as commerce, entertainment, leisure and information resources.

Copyright © 2003, Idea Group Inc. Copying or distributing in print or electronic forms without written permission of Idea Group Inc. is prohibited.

The first e-commerce applications were started in the early 1970s. The original applications were in the form of electronic fund transfers (EFT). These applications were limited to larger corporations and financial institutions (Turban et al., 2000). This type of transaction later included electronic data interchange (EDI). There is a marked difference between EDI and e-commerce in that e-commerce involves

*Table 2: Definitions of E-Commerce*

| Author | Definition/Approach/Description |
|---|---|
| Mclaren and Mclaren, 2000 | Any electronic business transaction or exchange of information to conduct business. |
| Greenstein and Feinman, 2000 | E-commerce is the exchange of products and services that require transportation, via some form of telecommunication medium from one location to another. Electronic business is defined as the exchange of information and customer support. The activities supporting, for example, the exchange of information and provision of customer support are not strictly speaking 'commerce' activities, but can be referred to as 'business' activities. |
| Ford and Baum, 1997 | See e-commerce as an umbrella term that includes automated business transactions, online purchases, electronic forms and industrial inventory control transactions. They conclude the e-commerce represents a broad range of technologies, is socially accepted and is expected to be used. |
| Turban et al., 2001 | Business done online; the exchange of products, services and money with the support of computers and computer networks |
| Whitten et al., 2001 | Involves conducting both internal and external business over the Internet, intranet and extranets. Electronic commerce includes the buying and selling of goods and services, the transfer of funds and the simplification of day-to-day business processes - all through digital communication. |
| U.S. Department of Commerce, 1999 | Business processes which shift transactions to the Internet or some other non-proprietary, Web-based system. |
| Turban et al, 2000 | Electronic commerce is an emerging concept that describes the process of buying and selling or exchanging products, services and information via computer networks, including the Internet. |

Copyright © 2003, Idea Group Inc. Copying or distributing in print or electronic forms without written permission of Idea Group Inc. is prohibited.

much more than EDI (Greenstein & Feinman, 2000). There is no standard definition for e-commerce. Table 2 presents some definitions of e-commerce.

In principle, however, most authors agree that e-commerce uses some form of transmission medium through which an exchange of information takes place in order to conduct business (Barnard & Wesson, 2000).

E-commerce can be classified in different ways. Turban et al. (2000) provide the following classifications (see Table 3).

The authors suggest that, because of the different types of e-commerce, that a different set of accounting systems methods should be used to disclose the nature of the use of technology in the organization.

## AUDITABLE

This is the process of independently examining, evaluating and advising on systems and their control so that, firstly, the information stored or output produced

*Table 3: Classifications of E-Commerce*

| Classification | Description |
|---|---|
| Business-to-Business (B2B) | This includes inter-organizational information systems and electronic transactions between organizations. |
| Business-to-Consumer (B2C) | B2C transactions are mostly retailing transactions with individual customers or consumers. |
| Consumer-to-Consumer (C2C) | C2C involves consumers selling directly to other consumers. This type of application includes auction sites and advertising personal services on the Internet. It can also include intranets and other organizational networks to advertise items and services. |
| Consumer-to-Business (C2B) | In this category, one will find consumers who sell to organizations. It also includes individuals who seek sellers with whom they may interact in order to conclude a transaction. |
| Non-business E-Commerce | Many institutions or organizations also use e-commerce to improve their operation and their customer services. |
| Intrabusiness (organizational) E-Commerce | All internal organizational activities involving exchange of goods, services or information usually performed on intranets are included in this category. |

*Source: Adapted from Turban et al. (2000)*

Copyright © 2003, Idea Group Inc. Copying or distributing in print or electronic forms without written permission of Idea Group Inc. is prohibited.

shall conform to some externally required specification and secondly, that its continued integrity and reliability is assured (British Computer Society, 1980).

An auditor should be able to retrieve a set of records associated with a given entity and determine that those records contain the truth, the whole truth, and nothing but the truth. There should be a reasonable probability that any attempt to record incorrect, incomplete or extra information will be detected. Thus, even though many transactions will never be scrutinized, the falsification of records is discouraged. This section categorizes the many ways in which records can be falsified.

Entity X has a transaction with an arbitrary number of other parties. Each should then record the agreed-upon transaction T in their records. Records can lose integrity as follows:

- X does not record transaction T.
  1. All parties in the transaction agree not to record transaction T.
  2. X does not record the transaction, but one party, Y, does. X and Y's records are in conflict.
  3. X used a false identity in the transaction T so that its records would not directly contradict those of Y.
- X inserts an incorrect transaction record T in its records at the time of the transaction instead of T.
  4. All parties in the transaction agree to record the false transaction T.
  5. X records false information T, but some parties record T.
- X invents and records a transaction that did not occur.
  6. All parties in the alleged transaction agree to support X's falsification.
  7. X records the false transaction without support from all other parties.
- X inserts incorrect information into the records after the fact. (This option is often particularly beneficial to perpetrators.)
  8. X alters transaction T after it has been recorded.
  9. X removes a prior transaction T from the records.
  10. At the time of transaction T, X records multiple versions. X later selectively forgets all versions but one.
  11. All parties agree to delay recording a transaction. They subsequently select false information to record, or decide not to record the transaction at all.
  12. X retroactively records a transaction that never occurred.

Of the 12 approaches described above, three require all parties involved in a transaction (or alleged transaction) to conspire to falsify records at the time of the transaction (1, 4, and 6). No system can possibly detect such a conspiracy, whether

Copyright © 2003, Idea Group Inc. Copying or distributing in print or electronic forms without written permission of Idea Group Inc. is prohibited.

the transaction takes place over a telecommunications network or in person. For example, if a storeowner and a customer agree to enter an incorrect price on a cash register, records will show no signs of falsification. However, in every other case listed above, an inconsistency can potentially be observed. A system for electronic transactions should be as auditable as transactions in the physical world. Thus, in an auditable system, every other case should be detected, and it should be possible to identify the guilty parties. *This should be possible even if the parties in a transaction and operators of the system itself cooperate to falsify records.* (Peha, 1999)

# SOUTH AFRICAN REVENUE SERVICES (SARS)

In an effort to streamline the tax collection process, the SARS has introduced e-filing.

E-filing will be facilitated through service providers, external companies with the necessary infrastructure to provide electronic submission services. Taxpayers who wish to file and pay electronically will register with the Service Provider of their choice, conclude an agreement and receive a private access code and password to access the available services. The private access code and password will only be issued once the service provider has authenticated the taxpayer. The service provider will forward the necessary details of the taxpayer to SARS in order for the taxpayer to be activated as an e-filer on the SARS systems. When returns are to be issued, SARS will issue the electronic returns to the Service Provider with whom the taxpayer is registered. The Service Provider will in turn issue a reminder to the taxpayer, either by SMS or e-mail, informing him/her of the returns that have been received from SARS. The taxpayer will utilize his private access code and password to access the return. The Web-based application will automatically display the return information as received from SARS. The taxpayer will then complete the return on the Web. When the taxpayer electronically completes the return, *the details entered onto the return are validated thoroughly and all calculations are performed by the system in order to eliminate any potential of transmitting incorrect information.* The taxpayer also has the option to make a payment when submitting a return and can make the payment any time before the due date (SARS, 2002).

This three-way communication (between taxpayer, service provider and SARS) could be fraught with all types of technical and non-technical problems. For example, who is responsible for the accuracy and integrity of the data? Who is responsible for the accuracy and integrity of the data warehouse, and how would you data mine seemingly different sets of data?

Copyright © 2003, Idea Group Inc. Copying or distributing in print or electronic forms without written permission of Idea Group Inc. is prohibited.

# LEGACY

Turban et al. (1999) define legacy systems as older, usually mature, information systems. Some have been around for up to 30 or 40 years. Some are less than ten years old. They are often mainframe or distributed systems in which PCs act as smart terminals. Legacy systems may include LANs and even some relatively recent client/server implementations.

Moore's Law (1965) suggests that the processing power of computers doubles every 18 months. Moore has also applied this law to the Web and E-commerce, immediately introducing the concept of legacy in such systems.

The rapid advances in hardware and the expansion of the WWW have now added a new dimension to legacy applications. An *entire CBIS* may join the ranks of a legacy application within one to one-and-a half years of its development. This poses a special problem for computer auditing. Somehow, an internal auditor has to 'freeze' the state of the technology at the time that the business processes were in operation, retaining knowledge of how the particular activities were dealt with by technologies that were useful at some time in the past but are no longer applicable. The business organization may need to preserve both the data and the processing methods for post-facto inspection by an internal auditor.

# POSSIBLE PROBLEMS

Many organizations in South Africa now offer many services on the Internet, such as credit-card purchasing, electronic bill-payment services and digital cash. The emergence of digital cash has transformed the 'accounting/auditing equation' and introduced many new aspects of security and integrity. Computer security involves the maintenance of three characteristics: confidentiality, integrity and availability.

*Table 4: Adapted from Pfleeger, 1997*

| Security Goal | Description |
|---|---|
| Confidentiality | Means that the assets of a computing system are accessible only by authorized parties. This type of access is read-type access: reading, viewing, printing or even just knowing about the existence of an object. |
| Integrity | Means that assets can be modified only by authorized parties or only in authorized ways. In this context, modification includes writing, changing, changing status, deleting and creating. |
| Availability | Means that assets are accessible to authorized parties. An authorized party should not be prevented from accessing objects to which they have legitimate access. |

Copyright © 2003, Idea Group Inc. Copying or distributing in print or electronic forms without written permission of Idea Group Inc. is prohibited.

Wilkinson (1989) identifies three major approaches to audits of computer-based systems. These are auditing around the computer, auditing through the computer and auditing with the computer. The first two approaches are mainly used for tests of controls; they involve such techniques as traces of selected transactions, test data, integrated test facilities, parallel simulations and embedded audit modules. The auditing with the computer approach uses the Generalised Audit Software (GAS) technique. These audits are now performed in a rapidly changing technological environment, and the emergence of the WWW and the e-phenomenon necessitate a modification of the audit procedure for electronic transactions to suit that environment.

The Internet creates many more possibilities for improving life. For example, classified ads bring in a large percentage of newspaper industry revenues, but ads can be replaced by a much cheaper and more convenient electronic system. It is often thought that instead of buying an entire newspaper, readers will pay for the individual stories in which they are interested. Someone wishing to purchase a VCR might send an 'intelligent agent' into the Internet to collect bids from suppliers for a unit that meets desired specifications, and then selects the best choice. While such scenarios are technically feasible, some industries may not allow software agents to collect their prices (Odlyzko, 1997).

Many artificial barriers could be introduced into this digital economy. For example, in 1990, IBM introduced the LaserPrinter E, a lower cost version of its LaserPrinter. The two versions were identical, except that the E version printed five pages per minute instead of ten for the regular one. This was achieved (as was found by independent testers, and was not advertised by IBM) through the addition of additional chips to the E version that did nothing but slow down processing. Thus, the E model cost more to produce, sold for less and was less useful. Consumers who do not need to print much, and are not willing to pay for the more expensive version, obtained this laser printer. Consumers who do need high capacity obtain a lower price than they might otherwise have to pay, since the manufacturer's fixed costs are spread over more units (Odlyzko, 1997). Another artificial barrier that could possibly lead to legacy is the bundling of software. Bundling consists of several software goods packaged together, such as a word processor, a spreadsheet and a presentation program in one software suite (such as Microsoft Office) (Odlyzko, 1997). Because of this marketing strategy, some of these products become the *de facto* standard and even when there is a move away from these products, we still have people using them.

Copyright © 2003, Idea Group Inc. Copying or distributing in print or electronic forms without written permission of Idea Group Inc. is prohibited.

# E-CAP

The Johannesburg Stock Exchange (JSE) Securities Exchange has signed an agreement with the London Stock Exchange that will result in local stocks being traded electronically via London. This will encourage South African companies to keep their primary listing in SA, by allowing them to have dual primary listing in both countries (Ebersöhn, 2001). Such trading methods would naturally need and use government and legal regulations governing these transactions.

In a broad sense, accounting has the following hierarchy, with the inclusion of e-commerce into this hierarchy suggested by the authors, as shown in Figure 1.

There are obvious overlaps between these areas.

Under e-commerce, the authors present the following subdivision with their related activities; see Figure 2 and Figure 3. Each subdivision would have a particular set of guidelines that aid the auditor.

*Figure 1*

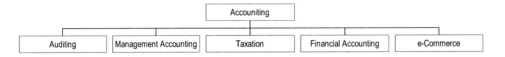

*Figure 2*

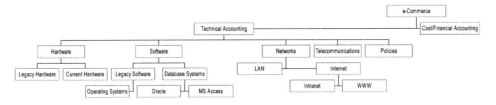

*Figure 3*

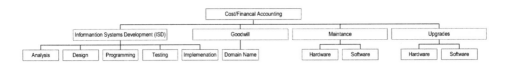

Copyright © 2003, Idea Group Inc. Copying or distributing in print or electronic forms without written permission of Idea Group Inc. is prohibited.

The traditional accounting methods do not deal with the unique complexities that e-commerce introduces into the business. Accounting standards do not specifically deal with the financial reporting standards for Internet firms, often giving these firms abundant freedom in their financial reporting. For example, the development cost of Web sites should be amortized evenly over their useful economic lives. Costs related to the development of a web site from planning, design, implementation and maintenance should be accounted for differently, e.g. infrastructure and initial set-up are to be capitalized. How do auditors measure the useful economic lives of Web sites, domain names, subscribers and goodwill (Coppin, 2001)?

As a sign of increasing usage of the Internet for business KPMG (2001) reports the following:

- Several companies highlighted the use of the Internet in core insurance operations, within their financial statements.
- Companies are disclosing legal issues related to e-commerce on their financial statements.
- Some companies have integrated their comments on e-commerce into the Chairman's/Chief Executive's statement in order to emphasize that e-commerce is an integral part of their business activities rather than just a distribution tool.
- A number of insurers are using Internet technology to support their intermediaries.
- A major international assurance company offers its customers the ability to apply for mortgages and pensions on-line and to obtain valuations of their pensions and unit trusts.
- A major insurer has developed an Internet concept for retail consumers, which gives quotes for motor insurance after six questions. It selects the most competitive quote from a panel of insurers. It has also developed Internet joint ventures with a car manufacturer.

To approach some of these control issues above the authors propose the development of a framework called e-CAP (Electronic Commerce Accepted Practice). E-CAP's purpose is providing guidelines for systematically and properly dealing with the processing of electronic transactions.

Table 6 contains examples of some issues affecting electronic commerce.

The impact of the connectivity age is vast. Though it is early days for e-commerce, there is an urgent need for professional people (IT professionals,

Copyright © 2003, Idea Group Inc. Copying or distributing in print or electronic forms without written permission of Idea Group Inc. is prohibited.

*Table 5: Adapted from Turban, et al., 2001*

| Current Limitations | E-CAP guideline/policy statement |
| --- | --- |
| There is a lack of universally accepted standards for security and reliability. | E-CAP would provide advice relating to e-Commerce, networked environments (e.g., virtual private networks) without giving away trade secrets. |
| Software development tools are still evolving. | For example, cookies (W1, 2002): A message given to a Web browser by a Web server. The browser stores the message in a text file. The message is then sent back to the server each time the browser requests a page from the server. <br><br>The purpose of cookies is to identify users and possibly prepare customized Web pages for them. When you enter a Web site using cookies, you may be asked to fill out a form providing such information as your name and interests. This information is packaged into a cookie and sent to your Web browser, which stores it for later use. The next time you go to the same Web site, your browser will send the cookie to the Web server. The server can use this information to present you with custom Web pages. Therefore, for example, instead of seeing just a generic welcome page you might see a welcome page with your name on it. Content of cookies will need to be scrutinized by a verification authority. |
| There are difficulties in integrating the Internet and e-commerce software with existing (especially legacy) applications and databases. | E-CAP would provide levels of disclosure of the state of a company's technology without revealing trade secrets, whilst maintaining investor/stakeholder confidence. |
| Many legal issues are yet unresolved. | E-CAP would provide some prudent rules for auditors to deal with legal anomalies, such as digital signatures. |
| Web page risk profile | A defined series of icons that indicate the type of risk related to that industry, such that at one glance, auditors and final users can estimate the related risk. |
| <ul><li>Who is responsible for the authenticity, fidelity and accuracy of information?</li><li>How to ensure that information will be processed properly and presented accurately to users?</li><li>How to ensure that errors are accidental in databases, data transmissions and data processing and not intentional?</li><li>Who is to be held accountable for errors in information? How should the injured party be compensated? How is this accounted for in financial statements?</li></ul> | E-CAP would provide a guideline that would outline how to assess an online backup and recovery process, how to deal with sensitive data in log files, cache servers, and archives that collect cookies. |

Copyright © 2003, Idea Group Inc. Copying or distributing in print or electronic forms without written permission of Idea Group Inc. is prohibited.

accountants, auditors, government regulators, stakeholders and lawyers) to discuss and document the effect of e-commerce and provide preliminary documents for e-commerce transactions. Prudent accounting practices have to be adopted to avoid devious and/or inaccurate disclosures.

### Electronic Communications and Transactions Bill (Electronic Communications and Transactions Bill 2002)

The Electronic Communications and Transactions Bill is an attempt by the Republic of South Africa to provide for the facilitation and regulation of electronic communications and transactions. It intends to provide for the development of a national e-strategy for the Republic; to promote universal access to electronic communications and transactions and the use of electronic transactions by Small, Medium and Micro enterprises (SMMEs). It also intends to provide for human resource development in electronic transactions, to prevent abuse of information systems, to encourage the use of e-government services; and to provide for matters connected therewith.

Some of the objects of the Act are to enable and facilitate electronic communications and transactions in the public interest, and for that purpose to—

*(a)*   recognize the importance of the information economy for the economic and social prosperity of the Republic;

*(b)*   promote universal access primarily in underserviced areas;

*(c)*   promote the understanding and, acceptance of and growth in the number of electronic transactions in the Republic;

*(d)*   remove and prevent barriers to electronic communications and transactions in the Republic;

*(e)*   promote legal certainty and confidence in respect of electronic communications and transactions;

*(f)*   promote technology neutrality in the application of legislation to electronic communications and transactions;

*(g)*   promote e-government services and electronic communications and transactions with public and private bodies, institutions and citizens;

*(h)*   ensure that electronic transactions in the Republic conform to the highest international standards;

*(i)*   encourage investment and innovation in respect of electronic transactions in the Republic;

*(j)*   develop a safe, secure and effective environment for the consumer, business and the Government to conduct and use electronic transactions;

*(k)*   promote the development of electronic transactions services which are

Copyright © 2003, Idea Group Inc. Copying or distributing in print or electronic forms without written permission of Idea Group Inc. is prohibited.

responsive to the needs of users and consumers;

*(l)*    ensure that, in relation to the provision of electronic transactions services, the special needs of particular communities and, areas and the disabled are duly taken into account;

*(m)*    ensure compliance with accepted International technical standards in the provision and development of electronic communications and transactions;

*(n)*    promote the stability of electronic transactions in the Republic;

*(o)*    promote the development of human resources in the electronic transactions environment;

*(p)*    promote SMMEs within the electronic transactions environment;

*(q)*    ensure efficient use and management of the .za domain name space; and

*(r)*    ensure that the national interest of the Republic is not compromised through the use of electronic communications.

These are the first steps in developing a manageable framework for the sustainable development of electronic commerce in South Africa. Such a mammoth task requires a well-strategized, thoroughly planned and carefully coordinated approach to electronic commerce. Partnerships between Government and industry will be required, not only to develop the actual strategies, but also to become involved in the integration of the existing, future and newly created digital world entities. Without a cohesive outlook and attitude to such a challenge, the expected benefits may not accrue.

# E-CAP'S ABCD FRAMEWORK

The letters ABCD symbolize the grass roots approach: Starting at the beginning and not assuming anything about the audience.

*   A stands for *atmosphere*: the organization should understand the atmosphere in the particular environment that it is operating in, e.g., socio-political, law, local customs, and the local languages spoken.
*   B stands for *build-up*: an electronic entity has to build-up a culture of trust between itself and customers, who are in any part of the world.
*   C stands for *communication*: the build up of good communication lines between customers, suppliers and the organization. Being at a distance from the community has the effect of dehumanizing the relationship.
*   D stands for *discipline*: work within the rules and regulations of the community.

Copyright © 2003, Idea Group Inc. Copying or distributing in print or electronic forms without written permission of Idea Group Inc. is prohibited.

*Figure 4: Source: Wang, 1999*

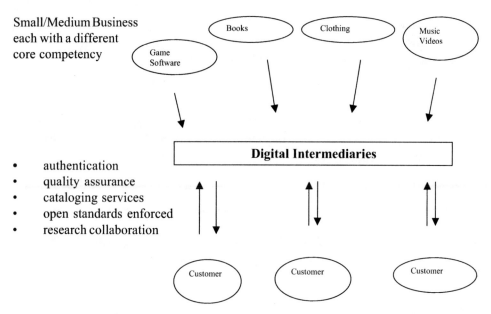

The ABCD approach is a simple approach that assists the company in assessing its social environment.

# FURTHER ISSUES TO CONSIDER

E-commerce will only be encouraged if the infrastructure supports the technology. An online customer may not feel adequately protected, choose to research a product then buy off-line. There may arise a "digital divide" in the business world: the larger, well-established companies *effectively* use the technology forcing small and medium businesses to adapt or die. Governmental regulation and legal issues would affect e-commerce, e.g., the question of when is a digital contact concluded? In June 2000, the President of the United States signed the Electronic Signatures in Global and National Commerce Act (E-Sign). This provides equal legal status between electronic and traditional signatures, given mutual consent between the parties to use E-Sign. Many countries will need to follow this path awaiting international standards (McNurlin & Spragne, 2002). Wang (1999) reports that 'medium and small business often do not fully understand the concept of information systems and its ability to facilitate electronic commerce.

Copyright © 2003, Idea Group Inc. Copying or distributing in print or electronic forms without written permission of Idea Group Inc. is prohibited.

Small firms are at times intimidated by the technology, are frequently concerned about the ability of outsiders to tap into the workings of a small firm via computers, and often lack time or resources to develop an understanding of how information systems help them'.

Technology alliances (Wang, 1999) may be formed between different business organizations as well as companies using Digital Intermediaries, e.g., see Figure 4.

All these issues affect the way an audit will be conducted. Crisis management should be avoided. Guidelines for dealing with e-commerce in a realistic and standard accounting approach are needed.

# REFERENCES

Barnard, L. & Wesson, J. (2000). E-Commerce: An Investigation into Usability Issues, in *Proceedings of the 2000 South Africa Institute of Computer Science and Information Technologists (SAICSIT)*. South Africa, Cape Town, November 1-3. Available at: http://www.cs.wits.ac.za/~philip/SAICSIT/SAICSIT-2000/programme/papers.html.

Bishop, D., Bishop, C. & Bishop, O. (1996). *An introduction to the World Wide Web for PC and Mac users*, London, England: Bernard Babani Ltd.

British Computer Society, (1980). *Control and Audit of Minicomputer Systems*, London, England: Heyden & Son Ltd.

Business 2.0 (2000). *Business 2.0: 10 Driving Principles of the New Economy*. MB Worksoft, Intelligence Supplement.

Coppin, G. (2001). *Financial reporting by Internet firms*. Accounting SA, April.

Ebersöhn F. (2001). *JSE launches London crossing*, Business Report, April 3.

Electronic Communications and Transactions Bill 8 of 2002, Republic of South Africa

Erwin, G.J & Singh, S (2001). The Legacy of the Web, *in Proceedings of the 2001 Information Resources Management Association International Conference (IRMA)*. Toronto, Ontario, Canada. May 20-23.

Ford, W. & Baum, S.M. (1997). *Secure Electronic Commerce*. New Jersey: Prentice-Hall.

Gilfillan, M. (2000). Turing E-business into Business-As-Usual, in *Business 2.0: 10 Driving Principles of the New Economy*. MB Worksoft, Intelligence Supplement.

Greenstein, M. & Feinman, T.M. (2000). *Electronic Commerce: Security, Risk Management and Control*. Boston, MA: Irwin McGraw-Hill.

Copyright © 2003, Idea Group Inc. Copying or distributing in print or electronic forms without written permission of Idea Group Inc. is prohibited.

KPMG (2001). *Principles and Presentation – Insurance, Survey of UK insurers' 1999 financial statements.* Available at: http://www.kpmg.co.uk/kpmg/uk/services/finsect/publics/ppins/ppins10.pdf.

McNurlin, C.B. & Sprague, R.H. (2002). *Information Systems Management in Practice,* 5th ed., New Jersey: Prentice-Hall.

McLaren, C.H & McLaren, B.J (2000). *E-Commerce Business on the Internet.* South-Western Educational Publishing, Cincinnati.

Moore, G.E. (1965). *Moore's Law.* [Online] Available at: http://www.intel.com/research/silicon/mooreslaw.htm.

Odlyzko, A. (1997). The Bumpy Road of Electronic Commerce, in *Proceedings of ED-MEDIA 97 & ED-TELECOM 97,* June 14-19, in Calgary, Alberta, Canada.

Peha, J.M., (1999). Electronic Commerce with Verifiable Audit Trails, in *Proceedings of the 9th Annual Conference of the Internet Society.* Available at: http://www.comms.uab.es/inet99/1h/1h_1.htm.

Pfleeger, P.C (1997). *Security in Computers,* 2nd ed. New Jersey: Prentice-Hall.

Ricks, D.A. (1993). *Blunders in International Business.* Cambridge, MA: Blackwell.

SARS. (2002). File PAYE and Skills Development Levies Online! File VAT and Diesel Refunds Online! Available at: http://www.sars.gov.za/commerce/.

Stair, M.R. (1992). *Principles of Information Systems: A Managerial Approach,* Boston: Boyd and Fraser.

Turban, E., Rainer, R.K & Potter, R.E. (2001). *Introduction to Information Technology,* NY: Wiley.

Turban, E., Lee, J., King, D. & Chung, M.H. (2000). *Electronic Commerce: A Managerial Perspective.* New Jersey: Prentice-Hall.

Turban, E., McLean, E. & Westherbe, J. (1999). *Information Technology for Management: Making Connections for Strategic Advantage,* 2nd ed. NY: Wiley.

U.S. Department of Commerce (1999). *The Emerging Digital Economy II.* Available at: http://www.ecommerce.gov/ede/chapter1.html.

Whitten, J.L., Bentley, L.D. & Dittman, K.C. (2001). *Systems Analysis and Design Methods.* Boston: McGraw-Hill Irwin.

Wang, Y. (1999). A General Analysis of the Impact of Electronic Commerce on Small Businesses and Entrepreneurships in the U.S. in *Proceedings of the 1999 South Africa Institute of Computer Science and Information Technologists (SAICSIT).* South Africa, Johannesburg, November 17-19. Available at: http://www.cs.wits.ac.za/~philip/SAICSIT/SAICSIT-99/electronic/exp/Wang.pdf.

Copyright © 2003, Idea Group Inc. Copying or distributing in print or electronic forms without written permission of Idea Group Inc. is prohibited.

Wilkinson, W.J. (1989). *Accounting Information Systems: Essential Concepts and Applications*, Wiley, USA.

W1 (2002). Cookies, in *Webopedia*, Available at: http://www.webopedia.com/TERM/c/cookie.html.

Copyright © 2003, Idea Group Inc. Copying or distributing in print or electronic forms without written permission of Idea Group Inc. is prohibited.

## Chapter V

# The Theory Behind the Economic Role of Managing the Strategic Alignment of Organizations While Creating New Markets

Sam Lubbe
Cape Technikon, South Africa

## ABSTRACT

*Over the past couple of years, the Internet has taken off and organizations will soon reap economic benefits on it. E-commerce will therefore hopefully emerge as an efficient yet effective mode of creating new markets although most managers still doubt the economic impact and profitability it has. Enabled by global telecommunication networks and the convergence of computing, telecom, entertainment and publishing industries, e-commerce is supplanting (maybe replacing) traditional commerce. In the process, it is creating new economic opportunities for today's businesses, creating new market structures. Managers of tomorrow must therefore understand what e-commerce is; how the approach to this concept will be; and how it will affect the economic position of the organization.*

Copyright © 2003, Idea Group Inc. Copying or distributing in print or electronic forms without written permission of Idea Group Inc. is prohibited.

*These questions are asked: What is the return on investment (ROI) on e-commerce? What is the effect of e-commerce on the strategic alignment of the organization? In addition, what is the economic effect of the strategic alignment on the organization? This paper explains the economic impact of e-commerce and how it can be used to create new markets and to improve the strategic alignment of the organization.*

# INTRODUCTION

Over the past couple of years, the Internet has grown a lot and it is possible that organizations will soon reap economic benefits from using it. E-commerce, as one of the 'products' of the Internet will therefore hopefully emerge as an efficient yet effective mode of conducting global commerce although most managers still doubt the economic impact and profitability it has. Enabled by global telecommunication networks and the convergence of computing, telecom, entertainment and publishing industries, e-commerce is supplanting (and maybe replacing) traditional commerce. In the process, it is creating new economic opportunities and challenges for today's businesses, creating new market structures, and changing the alignment of the organization. All organizations increasingly must focus on their delivery speed while facing an increasing degree of uncertainty. How do organizations keep economic growth up while still delivering business value?

The establishment of linkage between business and IT objectives has also consistently been reported as one of the key concerns of IS managers (Reich & Benbasat, 1996). They argued that there is firstly a need to clarify the nature of the linkage construct (socially and intellectually) and secondly to report on a project that was developed to test the social dimension of this linkage. It is important that all executives are involved during the establishment of the economic and social linkage because it creates a better understanding of each other's long-term visions and their self-reported rating of the linkage and alignment. Based on the data they collected, Reich & Benbasat argued that understanding of current objectives and shared vision for the utilization of IT are proposed as promising potential measures for short- and long-term economic aspects of the reported social dimension of linkage and alignment.

Traditional processes and management concepts that focus on optimisation, efficiency, predictability, control, rigor and process improvement are not flexible enough to create the applications and infrastructure organizations need to achieve economic success in these competitive times. The organization needs to be adaptable and needs to focus less on what traditional e-commerce developers says

Copyright © 2003, Idea Group Inc. Copying or distributing in print or electronic forms without written permission of Idea Group Inc. is prohibited.

it 'ought' to do and more on what is actually working in the aggressive e-commerce-driven environment. Managers of tomorrow must therefore understand what e-commerce is; how the approach to this concept will be; and how it will affect the economic and social aspects of the organization. This chapter will therefore address the importance of economics on the strategic alignment of organizations. The chapter will firstly look at the strategic dimension of e-commerce in the organization. Next, the article will address the issue of e-commerce forecasting, IT alignment and e-commerce, the economic effect and the measurement of alignment while keeping the effect of e-commerce in mind. The summary of the chapter will be last.

## APPROACH TO THIS CHAPTER

This chapter will use the interpretive approach to research as an alternative to the more traditional positivist approach. By using the interpretive approach, this chapter refers to such procedures as those associated with inferential statistics, hermeneutics and phenomenology. The framework proposed is to formulate a specific goal (economics, e-commerce and social aspects), subject to certain constraints. This approach will call for the active role of people in strengthening a truly collaborative research effort to maintain a peaceful co-existence between all the boundaries. Behaviours such as pointlessness, absurdness or confusion can play a role during the alignment of the strategies in the organization. The phenomenological approach will therefore help with the validity of the interpretation that this article will conduct.

## THE STRATEGIC DIMENSION OF THE ORGANIZATION

Being competitive in the next couple of years will depend on the effective application of e-commerce. Organizations need to take a strategic perspective to ensure that investment in e-commerce is contributing to the organization's business strategy. Increasingly, e-commerce is therefore being used by innovative organizations to facilitate the economic force of technology strategies now that re-engineering has lost some of its lustre. Strategic alignment could also be defined as the extent to which the e-commerce strategy supports, and is supported by the business strategy.

Economic advantage should be based on the capabilities of the organization. Organizational learning processes produce economic advantages (Sohn, 1999). The organizational learning processes obtain knowledge from e-commerce,

Copyright © 2003, Idea Group Inc. Copying or distributing in print or electronic forms without written permission of Idea Group Inc. is prohibited.

interpret, distribute to the organization and memorize the knowledge. In order to have economic advantage, most organizations should have unique resources and capabilities. These resources and capabilities could help with the alignment of the strategies of the organization. Organizational learning, therefore, provides the opportunities to improve the existing e-commerce and eventually help with better alignment of the strategies.

Whenever a positive social culture exists, employees are encouraged to learn and be ready to accept new information or strategies and this could eventually also help with better alignment of strategies. Ulrich and Lake (1990) as *cited* by Sohn noted that organizational capabilities have four components, that is shared mindset, management and human social resources practices, capacity for change and leadership. These capabilities have all the ability to affect alignment; they should be kept in mind when alignment is discussed. Sohn argues that e-commerce influences organizational learning and that each process of organizational learning is affected and adjusted by e-commerce. These adjustments would affect the alignment of all strategies; the executives should keep them in mind.

Organizations should also use a set of measurements that could be used for measuring key business goals like economic impact; the results should be used to quantify its performance and emphasize its role in orchestrating alignment with all the strategies. In an age when economics, cost-cutting, social responsibly and adding competitive value lie more squarely on the shoulders of the organization it is no longer enough to successfully deploy systems - they should also be part of the alignment of all the strategies. Using metrics enables e-commerce to spotlight its value to get additional funding for future technology projects and ensure that everything aligns. There had been many questions on whether e-commerce contributes economically to the business and how efficient the organization is in aligning all strategies. No one had been able to effectively answer these questions. The tricky part of this question is how to align all strategies or measuring the contribution e-commerce makes to the organization. The use of e-commerce in organizations is subject to various kinds of risk and this is part of the alignment of the strategies (Bandyopadhyay et al., 1999). As spending on e-commerce rises steeply, organizations become increasingly technology-dependent and consequently become economically highly vulnerable to the risks of e-commerce failure and eventual failure of the organization because there would be no alignment of strategies. Risk management of e-commerce is therefore one of the important issues facing executives today. There is a framework, as discussed by Bandyopadhyay, et al. that concentrates on the sequential linkage of the four components of risk management which composes its entire.

Copyright © 2003, Idea Group Inc. Copying or distributing in print or electronic forms without written permission of Idea Group Inc. is prohibited.

According to them, this approach is an improvement because it enables managers to move smoothly from one component to another by identifying and understanding the possible courses of action in the different steps. What these managers should keep in mind is that the four steps could affect the alignment of e-commerce, although some of the external threats such as social impact and disasters are difficult to control. Managers should acknowledge that these could affect alignment of the strategies. Therefore, one should expect risk-reducing measures will eventually enforce an alignment of strategies. The key to understanding strategic risks is dependent on the organization's ability to foresee long-term benefits from a new system, assess the resources and capabilities of its potential competitors, assess its own financial strength, and align its e-commerce strategy with its overall business strategy (Bandyopadhyay et al., 1999).

Organizational strategy provides the vision of where any organization needs to go. The problem is that many organizations focus too much on tactics and operations, while the real value of e-commerce is to open new opportunities such as economics and to enter new markets such as e-commerce to induce potential customers to use the organization's products. Organizations could directly derive e-commerce and operational objectives from the overall objectives, ensuring alignment of e-commerce and business strategies. One of the problems every organization faces is that objectives and strategies can change according to market dynamics such as economics and social changes.

At the strategic level, senior managers need a clear vision of the competitive impact of e-commerce and how it will affect the alignment of e-commerce. Business strategy provides the vision of where the organization wants to go. Most organizations therefore focus on e-commerce (new revenue opportunities) to help the organization becomes competitive and enter new markets. It should be possible to ensure that technology objectives and operational objectives are directly derived from the organization's corporate objectives, ensuring that business and e-commerce organizations are focused on the goals and that they are aligned. Managers must keep in mind that a life cycle will provide a perspective of the formulation of these strategies because each phase of the life cycle has distinct characteristics that affect the operation of the business. Each one of these phases could 'experience' a gap and managers should ensure that economic advantage does not make these gaps bigger during any of these phases. These performance gaps need managing; the organization should determine its overall market posture considering its relative position in the industry.

To forecast the incremental advantage effect likely to result from implementing alternative alignment planning, organizations should structure profile measures concentrating on industry market potential, relevant industry sales and real market

Copyright © 2003, Idea Group Inc. Copying or distributing in print or electronic forms without written permission of Idea Group Inc. is prohibited.

share. This is where e-commerce should ensure that everyone who might use the product is using it as fully and as often as possible. The problem, however, is that visitor's numbers as claimed by Internet Service Providers (ISP) cannot be verified and until that can be done, will CEOs surely look at the economic and social impact as things that cannot be reconciled.

These gaps, as mentioned before, can contribute positively or negatively towards the economic strategy of the organization and it can affect the actual performance of the organization on the e-commerce strategy of the organization. These gaps are:

- **Product line gap**: Introducing improved or new products should ensure that the organization could compete on the Internet.
- **Distribution gap**: This is where e-commerce can help to expand the coverage, intensity and exposure of distribution. This can help to better align the IT strategy and that of the organization.
- **Usage gap:** The Internet should help to induce current users to try the product and encourage users to increase their usage.
- **Competitive gap:** This is where e-commerce and the economic and social impact approach of the organization can make inroads into the market position of competitors as well as product substitutes.

The future marketing approach can be used as an advantage by the organization and can be increased. The alignment of the organization's strategies can also be improved by increasing the organization's industry economic potential, which can be done by increasing relevant industry sales while maintaining the present market share or by improving the organization's real market share.

# IMPLICATIONS FOR IT BUSINESS VALUE DURING THE ALIGNMENT OF IS AND BUSINESS STRATEGY

Tallon and Kraemer (1998) noted the argument that organizations' inability to realize sufficient economic value from their e-commerce investment is because of an absence of strategic alignment. They cited Child (1992), who argued that the content of alignment should be a series on intersecting and mutually consistent choices across domains such as economic factors, business strategy, e-commerce strategy, organizational infrastructure and processes and e-commerce infrastructure and economic processes. Other authors noted that these domains do not allow considerations of strategic alignment as a continuous process, nor does it consider

Copyright © 2003, Idea Group Inc. Copying or distributing in print or electronic forms without written permission of Idea Group Inc. is prohibited.

management practices used in moving an organization towards alignment. There are tools available to oversee and manage the content and process of alignment.

According to Tallon and Kramer (1998), there are a number of benefits associated with process-level measures of strategic alignment. Process level measures are likely to yield greater insights into where the organization is misaligned, helping to isolate bottlenecks and other impediments to economic business value within the organization. If strategic alignment were measured at the organization level, e-commerce and business managers might simply know that their organization was misaligned. However, they would not have sufficient information to isolate the source of the misalignment. It would be somehow different if the organization adopted a process-level economic and social perspective and the strategy presented as a series of activities within each business process. Strategy is a series of intersecting activities, meaning that it fits neatly with the definition of any processes as a sequence or ordered set of activities. This would indicate that the organization should avoid having to force-fit strategy into one of the established generic strategy types and this force-fit strategy could have a similar effect on the economic strategy that organizations have to adopt. Measuring e-commerce and business strategy at the process-level allows organizations to take a closer look at key activities within each process configuration and to look at the e-commerce that support those activities.

Opportunities for strategic alignment will arise if technological resources are used for maintenance, improvement and creation of capabilities that underlie the business strategy. Therefore, resources and capabilities need to have a link between them. Strategic alignment is not an event, but a process of continuous adaptation and change. Therefore, the assignment of e-commerce resources to capabilities must be continuously re-evaluated to prevent the organization slipping into a state of misalignment. Organizations should also keep in mind that the ever-increasing pace of industrial, social, political, economic and environmental change underscores the importance of strategic alignment. E-commerce resources should thus be used to the maximum, ensuring effective use of e-commerce. This is a challenge that managers should keep in mind.

Tallon *et al.* (1999), find a relationship between economic value and strategic alignment in a sense that an absence of strategic alignment can lead to reduced payoffs from e-commerce investment. Their analysis also found that the e-commerce segment plays a key role in enabling an organization to convert strategic alignment into higher levels of e-commerce economic value. As organizations focus their efforts on achieving intangible impacts in areas such as innovation (for example, e-commerce) and customer relations (social impacts), evaluating these impacts should become a priority.

Copyright © 2003, Idea Group Inc. Copying or distributing in print or electronic forms without written permission of Idea Group Inc. is prohibited.

# ECONOMIC AND SOCIAL INFORMATION AND THE INTERNET'S INFLUENCE ON ALIGNMENT

Companies incessantly produce and use economic and social information in part because they perceive the information as a source of development. Industries and organizations need accurate and up-to-date economic and social information about companies and their financial performance. They also need reports on political, economic, social and market trends embracing environments such as manufacturing, wholesale and retail, government. Many people in different organizations can use this economic and social information. The information is important for things such as the operation of the organization and alignment of strategies. It is important to note that people should have adequate sources of information (internally as well as external information). In times of economic stringency, information services are regarded as the least essential arm of the organization. It should be remembered that the magnitude and complexity of the business market is largely unknown to the average customer and very few businesses have the machinery to collect and coordinate information from a variety of sources and to apply it towards improving organizational decision making and the alignment of strategies.

# THE IMPORTANCE OF DECISIONS AND DECISION FRAMEWORKS ON STRATEGIC ALIGNMENT

A decision is a position, an opinion or a judgement reached after consideration, according to the dictionary (Oxford Dictionary and Thesaurus, 1996). It is of importance to note that the definition does not state anything about logical or illogical economic and social analysis (that some organizations do while making decisions about alignment) or about good or bad results. Decisions are either good or bad until put within the context of an economic and social decision framework on strategic alignment (Cutter, 1999).

The framework on strategic alignment should be an organized sequence of decisions that managers must make. Every decision should have a demarcation of the technical terms that could help outline the economic and social decision and managers should weigh organizational and technical issues (in other words, issues that affect alignment) and make choices about products, e-commerce investment, economic processes and resources and social issues. Without a context, decisions are meaningless - for example, to invest in newer models of hardware means nothing

Copyright © 2003, Idea Group Inc. Copying or distributing in print or electronic forms without written permission of Idea Group Inc. is prohibited.

without understanding the framework within which the alignment decision was made. To understand the decision framework for strategic alignment is a prerequisite to delivering successful alignment. The decision making process would affect successful alignment of strategies and would have to include answers to questions such as (Highsmith, 1999):

- What types of economic decisions need to be made?
- Who (person or group) makes each type of decision and how does it affect the social issues involved?
- How does the organization create sustainable economic decisions?

The decision framework for strategic alignment should consist of contents, context and process and should be followed closely. The content part is the information and knowledge while the context part is the circumstances such as e-commerce, economic and social impact and other events. Process defines how participants during the strategic alignment process arrive at the decision of approaching the alignment. The decision to purchase a software product with known defects is an event that could affect strategic alignment. The circumstances (all e-commerce investment plans such as quality goals and competition) provide the context that shapes the decision to approach the alignment of all strategies.

# IT FORECASTING PLANS FOR ECONOMIC AND SOCIAL OBSOLESCENCE

It is hard to make honest forecasts without offending both those who prepare them and those who receive them. The same is true about the alignment of the strategies of the organization. As the organization approach a wave of alignment, its time for honest estimates about the economic returns that organizations can get from different uses of scarce resources - financially and human - in the next decade. Short-term alignments might work out because it is easy to suppress the knowledge that people affect the marketing of goods on the Internet. On the other hand, when organizations align these strategies on the longer term, organizations tend to under-predict the rate and extent of economic change because of the gap that exists. Surrounded by in-place e-commerce assets that represent gruelling effort and sizeable investment and endless sleepless nights to align the strategies, few of the organizations are daring enough to say that the pioneering alignment of strategies today could be tomorrow's obsolete trash and the cycle will have to start all over again.

Copyright © 2003, Idea Group Inc. Copying or distributing in print or electronic forms without written permission of Idea Group Inc. is prohibited.

Bloodgood and Salisbury (1999) claim that organizations may need to change the manner in which they operate. A change could eventually affect alignment. According to Bloodgood and Salisbury, there are two characteristics of the degree of knowledge creation, transfer and protection and the degree of tacitness of the organization's knowledge. These factors are considered the most influential in determining success of e-commerce in facilitating strategic change. Remember that any change in strategy will also facilitate a change in the alignment of the organization's strategy. Any changes in the strategy for e-commerce should keep corporate strategy in mind.

Organizations that use a knowledge creation strategy focus on economic creativity and experimentation to construct new knowledge for developing new products and services. New services such as the social impact of the Internet can and will affect the alignment of the strategies and should help with competitive advantage. Bloodgood and Salisbury state that to understand tacit knowledge is important to understand how strategic change may be accomplished and supported by e-commerce. They claim that reconfiguring with new resources combines both tacit and explicit knowledge in an effort to create something new and somewhat difficult to imitate. This new and difficult-to-imitate creation would have to be in alignment; this is what managers have to keep this in mind while planning the knowledge creation.

Long-range economic alignment and forecasts need unbiased evaluation of the use of technologies in actions such as e-commerce, but users must remember that everything will change drastically during the next three years. If organizations do not align their strategies, the alternative is to go bust. The question that they must ask before they align the strategies is, "What would they do if they were starting from scratch?" Competitors might be doing just that and they can compare at the same time what influence the e-commerce aspect of business could have on the organization.

When it comes to putting in new business computing systems and applications such as e-commerce, "progress" is not measurable in terms of distance from where the organization started. This measurement would also measure how far the organization has gone along the economic road to align the strategies unless the organization cannot see where they are going. This 'blind' travelling could be because the organization did not align all their economic and social strategies. Well-planned journeys (alignment) make provisions for measuring how aligned the strategies are and how far the organization could economically go before the alignment is unsound and the resources would have to be monitored, measuring the effect of e-commerce. This measurement could be difficult because nothing is sure about Internet traffic.

Copyright © 2003, Idea Group Inc. Copying or distributing in print or electronic forms without written permission of Idea Group Inc. is prohibited.

The developments ahead include many opportunities for organizations and as such create chances for development managers to deliver unprecedented economic value to the organizations. There are new social practices world wide that organizations could embrace in a number of ways. Tools such as Lean Development and Leadership-Collaboration Management could be used. These all should challenge the manager to question traditional ways and ensure that the manager create an e-commerce environment in which the organization can thrive.

BPR has lost some of the impetus and therefore e-commerce and organizational executives are again concentrating on aligning strategies (technology, economic, social and corporate). Organization still struggle to align these - especially to match them all and it seems most of the times as if the two are on opposite ends of the string. The e-commerce budget is like a hole, but in no way is it ever going to be filled and also getting the right people for the e-commerce initiative is a problem. People are always flying off pointing fingers at the e-commerce segment and this could affect the alignment and planning of the e-commerce strategies and business strategies. Most of the time, this could cause obsolescence because participating organizations do not always realize that achieving alignment involves personal dynamics and some physical dynamics.

# ORGANIZATION-E-COMMERCE ALIGNMENT AND ECONOMIC AND SOCIAL IMPACTS

Aligning business and e-commerce strategies continues to be an important management issue. The problem is that, according to the Cutter Consortium (1999), 65 percent of organizations have no e-commerce strategy, while 25 percent of organizations have not yet developed an overall e-commerce strategy. Only 4.2 percent of the organizations contacted *do* have an e-commerce strategy. It is obvious from their figures that most organizations do not take e-commerce seriously. Once the effect has hit these organizations, it would affect their alignment decisions. The Cutter Consortium claimed that many organizations have, however, noted that the role that the Internet can play, but they have not taken the steps necessary to fully realize its economic potential and again this would affect the alignment of their strategies should they try to realize these effects. E-commerce will affect the alignment of the organization and that people do not know what the initial and eventual affect will be. Technologies, however, affects all organizations - even the smallest ones. While being able to pose an opportunity, e-commerce also pose a threat to the organizations and managers need to recognize all the factors (e.g., economic and social impacts) that affect e-commerce and e-commerce alignment.

Copyright © 2003, Idea Group Inc. Copying or distributing in print or electronic forms without written permission of Idea Group Inc. is prohibited.

There is however, no framework for organizations to align strategies while conducting e-commerce. Some of the impediments on this alignment are: download delays, limitations in the interface, inadequate measurements of Internet traffic and successes, economic planning, security weaknesses and lack of standards on the Internet. On the other hand, the results for the organization could be positive in the fact that the organization experiences improved usage of the effectiveness of the organization's economic resources and improved return on the economic investment in data, software applications, technology and IT staff - in other words, improved quality.

The Internet is promoted as the essential way of doing business lately, but it is still a retail medium (Whiteley, 1999). Organizations that establish successful Internet economic operations will need effective alignment and good logistics on the supply side of their operations. Organizations should remember that e-commerce is commerce enabled by the Internet-era technologies. These technologies are:
- Electronic markets
- Electronic data interchange
- Internet commerce and analogous public ICT systems

In theory, the use of the Internet should give the consumer the opportunity to bypass the intermediary and, with appropriate interfaces directly affect alignment. Any of the e-commerce operations should affect the alignment of all strategies and the organization or the people working in the organization should not resist them. Organizations trading on the Internet, if they are able to build up a substantial business operation and align their strategies, will need to be slick in all economic and social aspects of their business. Organizations with wide ambitions would also have to plan carefully in order to align all their strategies. Internet commerce, however, is the great economic leveller, as the size of the organization would not affect their e-commerce operations. Therefore, if the organization grows, they will have to remember to use a good decision framework as a backup to ensure fast and efficient alignment of the strategies that should be a long-term economic investment.

On the other hand, many development teams work hard on projects, only to hear from their customers that it is not enough, priorities are incorrect or the economic results were wrong; these reactions could affect alignment. Technology changes and developments have been extraordinary in the last couple of years and all of these could affect alignment. The problem with this is that there seems to be less emphasis on business trends and solutions and more on technology and this could affect alignment. Highsmith (1999) argues that in the next decade, economic and social development strategies will be among the important strategy any

Copyright © 2003, Idea Group Inc. Copying or distributing in print or electronic forms without written permission of Idea Group Inc. is prohibited.

corporation will make. He also noted that sometime in the 1990s, software made the transition from an enabler of business processes to a driver of business economic strategies and thus would be helpful with alignment of e-commerce strategies. According to Highsmith, software is the new economy, yet the author reckons that all e-commerce should be taken into account when aligning all strategies.

Alignment should be taken with the appropriate decision and framework that suit the organization in the specific environment. As e-commerce capability evolves, it should enable processes, products and opportunities never considered (like economic and social ones). The first couple of months of the new millennium would take on frightening changes in the e-commerce environment, as we have never seen. To be able to handle this and to ensure that all technology and business matters can handle this, organizations should ensure better partnerships with their employees and the decisions frameworks they have designed.

To achieve the goal of being an aligned organization that ensures its proper alignment and *stays* aligned, they need to answer the following key questions:

- What business trends do e-commerce drive or with which it is co-evolving?
- What are the evolving key economic and social business and e-commerce strategies? How does the organization take advantage of them?
- What critical skills and capabilities assist the transition to the new economy?
- What necessary organization, infrastructure and management changes assist implementation of e-commerce and other strategies? How does the organization ensure their current and future alignment?
- How does management assemble a decision framework for the alignment of all e-commerce strategies? How do they communicate it to all the relevant people?

If the Internet (and this includes e-commerce) is a strategic business driver, how does the organization gives direction to this and what does the organization do to keep the economic advantage? Business alignment is about innovative ideas and putting these ideas into action more effectively than the same competitors. E-commerce should take a leadership role, but it depends on the organization and its culture and the types of strategic plans. Business trends such as market fragmentation, information capacity to treat customers on the Internet, shrinking product lifetimes and convergence of physical products and services are important ones and the organization need to concentrate more on these forces that are co-evolving with the e-commerce capability and opportunities that spring up.

Some strategies can help with alignment because they are highly involved with economic and social initiatives; e-commerce is knowledge creation and sharing,

Copyright © 2003, Idea Group Inc. Copying or distributing in print or electronic forms without written permission of Idea Group Inc. is prohibited.

collaboration, and agility. These are all overlapping and supportive of each other. Knowledge creation and sharing (as stated in this article) is the ability to create new knowledge from available information to help with alignment, while collaboration is the process of shared creation to draw on the expertise of participants and to be able to create alignment from this collaboration. Agility is the ability of the organization to use new knowledge to adapt to external stimulus and use this to better or ensure total alignment of all strategies. All of these would work better if the staff motivations were good enough and should be encouraged to ensure that there would be no atrophy.

Alignment is hard to see, but managers would know it. The combination of all these mentioned in the paragraphs above should ensure that alignment exists. Knowledge should also be there to ensure that there is alignment. The organization should also be sure to take all economic trends into account when aligning strategies. The question to ask is what skills will be necessary to align e-commerce and organizational strategies. However, if managers really understand what the organization is trying to deliver, then they should be able to align strategies. Systems thinking would therefore help with alignment, because it would ensure that the alignment and view of the entire organization as an economic and social unit.

On the other hand, reporting on user statistics on Internet sites is far below the required standards. These statistics could affect the alignment of strategies since no one can plan around uncertainty. That is why many CEOs have lost faith in the Internet as communication and business medium of the future. Many ISPs cannot tell you why the people have ended up on their site and how long they have stayed. The tracking technologies should solve the problem but nobody can guarantee anything - it is like TV - it is a guessing game and although the developers of tracking technology is aware of these problems, they had not found a solution to combat the problems. The problem is to discover what people do when they arrive on the site and if they ever return. This information is invaluable to managers and carries a lot of credibility. User tracking on the Internet therefore needs to mature and start to provide useful economic and social information those marketers and specialists and executives can use to align all the strategies of the corporation. The problem is that ISP's can include internal activities in their traffic or they can tamper with log files while the electronic counters are susceptible to server and network congestion and downtime. These facts would obviously affect Internet marketing and alignment of strategies.

Aligning the IS strategy with the organizations goals is also appropriate for those organizations in which there is a vision for items such as e-commerce. E-commerce is part of the actual choice any organization should have. This is where the E-commerce Department can play a major role - their primary responsibility

Copyright © 2003, Idea Group Inc. Copying or distributing in print or electronic forms without written permission of Idea Group Inc. is prohibited.

should be to create business opportunities (via e-commerce) and not merely to ensure that e-commerce exists for the whole organization. Their role would thus be to ensure that e-commerce strategy and organizational strategy align all of the time, while creating new opportunities.

White & Manning (1998) argued that the percentage of individuals on the Internet using the medium for shopping had increased by about 70 percent between 1995 and 1996. It is important to note that users expect to get something of value free and this type of marketing could affect the alignment of e-commerce strategy with the corporate strategy. Personal opinion should affect the alignment of the strategies. Gender appears to be important in e-commerce, which would have to be kept in mind during alignment of strategies. Consumer demographics on the Internet could differ from normal demographics (if ever); this deserves further research. The likelihood of purchasing from an online storefront would help with alignment on a new frontier and should be something that executives should keep in mind.

Everybody should row along, otherwise the technology would not be used to the best economic advantage of the organization. Everybody should thus use and learn about economic goals and this would make the alignment of corporate strategies easier. They would all sing along if everybody understands core business concerns. Everybody should also learn all topics or at least read about it - think about BPR, TQM, etc. The explosion of the Internet as an economic resource for business for business information in all countries have taken off and has made business people aware of how easily data can be stored in different locations and in different formats that all these topics can use. All this can cause growth and can affect strategic alignment. Some of the pitfalls are the following aspects:

- **Cost** can be very high, and unaffordable to organizations
- **Capacity development** lacks strategic capacities
- **Visibility and security** use other providers?
- **Inconsistency of telecommunications access** a problem in some areas since the Internet is not always available.

Executives should communicate the mission of e-commerce to the consumers, employees and other managers - this could win half their battle during alignment. It is also important to show other people how the E-commerce department operates so they understand problem handling. They can get feedback on the e-commerce economic role and services they render. They can check on end-users from time to time - even if they have no problems - just to make sure that everybody knows that your department exists. It is important that usable communication channels exist. The inclusion of users during and when a big project starts is also important.

Copyright © 2003, Idea Group Inc. Copying or distributing in print or electronic forms without written permission of Idea Group Inc. is prohibited.

Pay special attention to users with special needs and make sure that everybody receives the same amount of help. Make sure that the e-commerce department understand what applications people are using and why do they prefer to install their own software. All of these help with better alignment and overall functionality.

Gibson et al. (1998) note that with the onset of the 1990s came an unstable economic environment and businesses started to realize that the strategic vision of the business required a dramatic change in their economic and social environment in order to at least maintain their competitive advantage. This meant that everyone had to realign their strategies to stay competitive. According to them, the companies they investigated exhibited a change in strategic vision, but these changes were unreflected in their e-commerce. This could create a problem, because if e-commerce cannot reflect the changes in the economic strategic vision, it would affect the alignment of instruments that could be used to determine if the organization's strategies are aligned. It could be because certain economic and social changes could prevent the advent of developments such as e-commerce. Therefore, a gap exists which could create a misalignment between the organization's strategic vision and the route it follows. . Business and e-commerce create the gap, in this instance. Many of the organizations they investigated acknowledged that e-commerce was not a good fit for their organization. The companies that they investigated used a variety of approaches to realign the economic model and e-commerce with their strategic vision. Gibson et al. (1998) argued that these approaches have gone beyond technical solutions. Although they would eliminate the technical problems, they would probably not address the business issues, so realignment would not occur. New technology such as ERP software, combined with business process creates the need to realign the business model and e-commerce with their strategic vision.

The aim of management should be to try to provide insights into identifying areas that help or hinder the alignment of the strategy for the organization with the e-commerce. Alignment focuses on the activities that management performs to achieve cohesive goals across the organization. Certain activities can also assist in the achievement of the alignment while others are clearly barriers. Achieving this alignment is a dynamic and evolutionary process. The alignment requires strong support from management at all levels, good relationships within the organization, strong leadership. These points must be emphasized, heeded, and communicated well, so that all people understand the organization's business environment, at the same time ensuring a successful alignment. All enablers and inhibitors should be focused on ensuring successful alignment and therefore all inhibitors should be minimized.

Copyright © 2003, Idea Group Inc. Copying or distributing in print or electronic forms without written permission of Idea Group Inc. is prohibited.

# THE EFFECT OF ECONOMIC AND SOCIAL IMPLICATIONS DURING THE ALIGNMENT OF STRATEGIES

Organizations have to assume that there are some basic definitions that ensure that all terminology is clearly understood: There are two types of implications that managers can use: economic and social; there is a connection to these factors. Some basic definitions that can be used are:

- **Asset structure** - Assets the organization wants to finance or funds that they want to use to expand during e-commerce.
- **Financial structure** - Available funds or possible financing available to the organization.
- **Financial leverage** - The ratio of funds from outside the organization to the total assets of the organization. The act of borrowing is said to create financial leverage.
- **Operating leverage** - Refers to the extent to which total operating costs vary with changes in the operating revenues and these revenues could be affected with e-commerce.
- **Business risk** - Financial leverage increases risk because it makes the return realized by the investor more sensitive to any event affecting the performance or the asset purchased.

This chapter looks at the possible economic and social impact on alignment during e-commerce investment and expansion of the e-commerce operation. IT assets and the finance of the IT assets as demonstrated could be financed by the use of equity and of foreign economic financing. Leverage is a ratio that is calculated and shows any possible investor or manager (especially the e-commerce manager or e-commerce director) how a possible opportunity to obtain funds can have any effect of the organization and on alignment as the decision would be a strategy that would affect any other strategy.

Financial leverage affects analysis of interest and liquidity because of fixed commitments due to economics of funds from outside the organization. If the return on assets were higher than the cost of the debt, management may find that leverage has a positive effect. Leverage can increase the return on owner's equity but there is a risk factor that managers have to keep in mind. Financial leverage involves the use of funds obtained at a fixed cost in the hope of increasing the return to the owners of the organization. If funds are thus being obtained to help with e-commerce, managers have to keep in mind that a successful project could increase the return

Copyright © 2003, Idea Group Inc. Copying or distributing in print or electronic forms without written permission of Idea Group Inc. is prohibited.

on the funds while a negative project will reduce the return on the funds and this would affect alignment.

Business leverage has to do with the ratio between fixed costs and variable costs. This would have an affect on the economics of the organization. Fixed costs should always be recovered and assets with a fixed base cost should be used in the hope that more profit could be generated. It would be wise for the organization to use financial leverage if profit is stable, otherwise if fixed cost represents the majority of the expenses, managers would find that profit is not stable. Obviously, if customer numbers cannot be guaranteed in e-commerce projects, managers would find that profit would not be stable. Organizations are warned not to use financial leverage if business leverage plays a role in organizational strategies. Managers should also remember that certain assets could affect fixed costs negatively. Also that both these leverages could be combined, for example, an economic risk could be combined with low financial risks and the other way around. The total risk of the organization could entail a swap between total risk and the expected return on any investment. There are some economic ratios that could help with leverage:

- **Total debt to total assets**. This displays the percentage total funds supplied by other people such as creditors. They prefer a low ratio, while the owners prefer a big ratio to increase turnover and to keep control of the organization.
- **Total interest earned**. This ratio tells the organization how much turnover could be lowered without affecting the payment of these interest amounts. This ratio is calculated by dividing gross profit by interest.
- **Fixed cost coverage**. This ratio shows how the organization can pay fixed costs (interest added to long-term debt). The ratio is planned by calculating a *total for profit before tax* and *tax and rent* and this total is then divided by the *total for interest and rent*.
- **Breakeven point** is that mark where the total for fixed costs and variable costs is equal to that of the total turnover.
- The **degree of financial leverage** is calculated by dividing earnings before interest and taxes by earnings before tax minus interest.

There are other ratios that could affect financial or business leverage but are too many to mention and would not be dealt with during the course of this chapter. It would be difficult with e-commerce to calculate the break-even amount of sales that the organization needs because nobody can state for sure what the fixed and variable costs would be. This will affect leverage and one can safely note that e-commerce would affect the alignment of the organization.

Breakeven analysis could be used in three ways, to modernize a program or if the organization wants to use a new market such as e-commerce, to study the

Copyright © 2003, Idea Group Inc. Copying or distributing in print or electronic forms without written permission of Idea Group Inc. is prohibited.

effect of extensive build-up of the firm such as globalization and new product decisions.

Of importance for the organization is the effect of e-commerce on the economic and social impact and on the alignment of the organization. The cost to appear in the global market on the Internet could be affected as follows:

- The Internet market is perfect and all 'clickers' could get the correct information, these shoppers and browsers act rationally and that no costs per transaction completed exists.
- Personal and manufactured leverages are perfect substitutes and they would affect alignment in equal amounts.
- Corporations and individuals can borrow money at the same rate.
- All organizations on the Internet can be placed in the same risk class so that all classes are homogenous.
- Since it will be difficult to control tax on the Internet it should be assumed that no tax would be levied.
- The average income of any organization can be represented by a variable that could be picked statistically.

These arguments could be the following:
- The total economic value of any organization on the Internet should rise with the use of leverage
- The turnover ratio on the economic book value of owner's equity should rise with the advent of leverage but should not affect debt of the organization on the Internet
- Leverage should not affect debt unless the organization uses debt to finance the e-commerce.

Certain conditions could be analysed with the help of leverage - such as economics. Because the capital-market is not complete, will it also affect the capital structure? If organizations make the wrong decisions about leverage and alignment, then the following could happen:
- Possible incorrect application of financing
- Possible loss of qualified staff
- Loss of suppliers
- Loss of sales and the liquidation of the organization
- Economic and financial problems
- Loss of market share
- Formal liquidation procedures.

Copyright © 2003, Idea Group Inc. Copying or distributing in print or electronic forms without written permission of Idea Group Inc. is prohibited.

Managers should keep leverage in mind while aligning of the strategies takes place because it could adversely affect the organization.

# ORGANIZATIONAL INFORMATION VISUALIZATION - THE USE OF THE DIFFERENT INSTRUMENTS TO MEASURE ALIGNMENT

Customers of the e-commerce have recognized that e-commerce and alignment of strategies are synonymous terms. These could affect measurement of alignment or value derived from the applications support process and eventual alignment. Historically, there have been a number of challenges in aligning corporate strategies and creating an objective measurement of alignment improvement over time. It is important that measurement focuses on business objectives, end-user participation and satisfaction, reaching and sustaining alignment, tracking and measuring factors (such as economic and social responsibility) that affect alignment and communicate on alignment improvement. While remaining focused on these reported items, the alignment team must also be able to measure and report on their accomplishments.

They should be able to meet the challenges of being able to objectively measure alignment as performed by the management team and at the same time facilitate a process of continuous improvement of the alignment of the strategies; in other words, there should be a process to measure the alignment. Managers need to have a metric available that allows them to measure value in terms of business output and the following is required:

- Good metrics of past and anticipated future Internet strategies
- Sound management
- Detailed and disciplined tracking
- An agreed approach for management
- Good internal communication to all role players.

Measurement should be adaptable to both traditional application support and to enterprise resource planning on alignments. For the alignment team, the measurement approach should provide a focus on the end product - alignment of all strategies (deliverables that provide value to the organization) rather than simply tracking the alignment of strategies. It also ensures management visibility, in that communications show results in terms of accomplishments and milestones met. It should also provide an effective measurement capability that should be objective and reflects measurement capacity and alignment tracking.

Copyright © 2003, Idea Group Inc. Copying or distributing in print or electronic forms without written permission of Idea Group Inc. is prohibited.

Kiani (1998) argues that the current decade has witnessed evolution in the media environment and indicated that e-commerce could grow in importance. The opportunities offered by this new environment are still unknown and it is this fact that organizations should keep in mind while aligning all strategies. Kiani suggests new concepts and models for marketers to be kept in mind while aligning strategies. Organizations should keep in mind that they all compete in two worlds - a physical world that people can see and a virtual world made of information. It is thus clear that information should play a role while alignment of strategies takes place. The two-way communication channel between consumer and corporation, Kiani suggests, should be incorporated during the alignment process. This communication should be more in the way of dialogue.

The methodology could be to create a database of economic transaction histories and this database should be incorporated during the alignment process by management. This database would be moving all the time and it would mean that the alignment process would be more difficult and moving all the time to keep track of factors that could affect alignment. The unit of measurement could be the value of each Internet customer to the organization. Marketing strategy will be measured by changes in the asset value of the customer's base over time. Alignment of the strategies should be more flexible as niches too small to be served profitability could become viable as marketing strategies improves.

Opportunities on the WWW are equal for all players - regardless of size and this would be affected by the information available to the consumers. Customers could thus help with the alignment of strategies and become a 'partner' of the organization. Organizations have earned the right to the digital relationship and they have to shift their alignment if they continuously enhance the value they offer consumers. Alignment of strategies should keep in mind that strategies on Internet marketing and e-commerce would be how to attract users, engage the users' interest and participation, learn about the preferences of consumers and ensure that there is interaction.

With the flood of data produced by today, organizational decision makers must do something to allow players to extract the correct knowledge from the available information. Recent advances on the Internet and visualization technologies provide many organizations the capability to start using human visual/spatial capabilities to solve abstract problems found in business and as such the e-commerce aspect of the Internet. This could allow the decision maker to separate the rubbish from the best that there is. To achieve this, organizations must use e-commerce well and increase the value provided to normal and Internet customers. The Internet will force organizations to evaluate how and when they should start to use the Internet to create additional business value to their organization. Some

Copyright © 2003, Idea Group Inc. Copying or distributing in print or electronic forms without written permission of Idea Group Inc. is prohibited.

organizations could rely on strategic use of enterprise-wide e-commerce to enhance its competitive position as an established supplier of the goods they are marketing normally and on the Internet.

Consultants provide many measurement services. The statement that could be made is, "if you cannot measure it, you cannot align it." Once the organization establishes some objectives and goals, e-commerce can facilitate them through a set of technology initiatives and once organizations have that strategic sense, managers can identify criteria as to how this project creates value to the organization and how it can be used to help with alignment.

Many organizations employ the balanced scorecard technique to measure the e-commerce overall success in an ongoing process. This scorecard gauges things such as internal stakeholders' satisfaction, measures the system and economic values, rates the value and quality of that work and uses social responsibility that measure and help with the alignment of the organization's strategies. The report card is broken into three categories: responsiveness, value and quality. The users rate things such as deliverables, establishment of timelines, accurately identified timelines and whether the solutions meet the expected ROI. This scorecard gives employees the opportunity for dialogue with business users if results fall short. The balanced scorecard report card can be used to link the alignment of the strategies, regardless of the investment that was done. Things such as the business outcomes and organization desires, ensuring that key business holders are involved and the issues around change management would affect alignment. It is therefore about setting realistic, tangible and clearly communicated goals, ensuring that the entire team is speaking the same language and this could help with alignment.

To understand the success better organizations can use the balance scorecard to develop a model of organizational performance that could emphasize the contribution of e-commerce and the Internet to different dimensions of the performance. Some organizations would be able to prove that e-commerce, when properly aligned with the strategies can contribute to the organizations overall success and market leadership. Computer Simulation and Human Thought could help because strategic advantage could be obtained through effective use of the natural strengths each has to offer to improve the quality of the decision-making on outcomes such as e-commerce.

## Instruments that could be used to determine if the Organization's Strategies are aligned

The important thing to remember is to determine if the organization has aligned the strategies; one of their tools could be a questionnaire. The questionnaire could

Copyright © 2003, Idea Group Inc. Copying or distributing in print or electronic forms without written permission of Idea Group Inc. is prohibited.

be scored as follows. Count each category and determine what the people think of the alignment of strategies. If there are more Not Sures or Disagrees than Agrees, then there might be a problem. The steps to be taken would ensure open communication lines. The next table can help determine this.

The questionnaire available could be used in conjunction with the balanced scorecard approach. The balanced scorecard should keep enterprises such as e-commerce in mind and could be as follows: There should be different categories, combining it closely with the questionnaire, such as customer perspective, innovation and learning perspective, internal business processes perspective and the financial perspective. All of these categories should be discussed or be part of a decision framework where the different topics could include the goals for each perspective, the measurement of each perspective, the metrics being used, the targets and the actual that could be measured. For customer perspective, organizations need to remember that this could help with better alignment but should not carry as much weight as the internal metrics.

The innovation and learning perspective should be important as this is where new stuff such as e-commerce would be added and this would affect the alignment of all strategies. All new innovations and possible innovations should be added here. The internal business perspective should be used, as we need to ensure that employees have kept track of all the perspectives and that they need to know how to work with data, information and knowledge. The financial perspective should concentrate on increased efficiency, effectiveness and transformation and investment. As different measurement issues exist for these, they should be placed and used as and where needed.

# SUMMARY

E-commerce will develop over the next couple of years on the Internet and it is possible that organizations will reap economic benefits from using it. As one of the 'products' of the Internet, e-commerce will therefore emerge as an efficient yet effective mode of conducting global commerce. The problem is, as mentioned in this article, that managers still doubt the economic impact and profitability it has. In the process it is creating new opportunities and challenges for today's businesses, creating new market structures, and changing the alignment of the organization. An increasing focus on delivery speed and an increasing degree of uncertainty challenges all organizations. The question: How do organizations keep up while still delivering business value was answered.

This article has addressed the importance of economics on the strategic alignment of organizations. The article also looked at the strategic dimension of e-

Copyright © 2003, Idea Group Inc. Copying or distributing in print or electronic forms without written permission of Idea Group Inc. is prohibited.

*Table 1: A Sample Questionnaire that can help determine the Level of Strategic Alignment*

| Questions | | Agree | Disagree | Not Sure |
|---|---|---|---|---|
| 1. | Executive management is involved in all strategic information decisions and reference the formal business and information strategic plans. | | | |
| 2. | Customers, users and industry are regularly surveyed regarding the information needs and problems related to doing business with the organization. | | | |
| 3. | Technology is invested into only after establishing a business use for the decisions. | | | |
| 4. | The decision-making and operational roles of executives, managers and users are overlapping. | | | |
| 5. | Executives, managers and users understand and practice the concepts of managing data at all levels. | | | |
| 6. | Executives, managers and users understand and practice the concepts of changing information into knowledge at all levels. | | | |
| 7. | Information Services uses graphical, easy to understand methods of explaining how knowledge supports the organization's strategy. | | | |
| 8. | IT can help explain to you how the organization uses frameworks to align all strategies | | | |
| 9. | More than half the IT projects are under budget and on time. | | | |
| 10. | You know exactly how many IT projects are currently being conducted. | | | |
| 11. | You know exactly why and how to use all information and knowledge available | | | |
| 12. | You know how to apply the decision framework (if available) that will be used for strategic alignment. | | | |

commerce in the organization and addressed the issue of e-commerce forecasting, e-commerce alignment and e-commerce, the leverage effect and the measurement of alignment while keeping the effect of economics in mind.

Organizations that survive and thrive in the new economy and use all to ensure that their strategies are aligned are those with better ideas than the rest. These organizations normally produce better products, novel interpretations of the market (e-commerce in this instance), innovative management strategies and the ability to create a unified strategic alignment where pools of talent, both inside and outside the organization, ensure that all stays on track and as the organization moves, the alignment moves to ensure that everything is in 'harmony'.

Copyright © 2003, Idea Group Inc. Copying or distributing in print or electronic forms without written permission of Idea Group Inc. is prohibited.

# REFERENCES

Bandyopadhyay K., Mykytyn P. & Mykytyn M. (1999). A framework for integrated risk management in information technology, *Management Decision 37 (/5).*

Bloodgood J.M & Salisbury W.D. (1998). If the IT strategy fits wear it: Matching strategic change efforts with IT efforts, *Proceedings of 1998 Americas Conference,* Baltimore, Maryland, August 14-16

Child. (1992) cited by 1998 in A Process-oriented Assessment of the Alignment of Information Systems and Business Strategy: Implications for Business Value, *Proceedings of 1998 Americas Conference,* Baltimore, Maryland, August 14-16

Cutter Consortium. (1999). Weekly e-mail service for IT professionals - Using IT work units (dated 14/09/99) and E-business software practices (dated 01/09/99).

Gibson N., Holland C. & Light B. (1998). Identwing Misalignment between Strategic Vision and Legacy Information Systems in Organizations, *Proceedings of 1998 Americas Conference,* Baltimore, Maryland, August 14-16

Highsmith J. (1999). *Thriving in Turbulent Times,* White paper by Cutter Consortium

Kiani G.R. (1998). Marketing opportunities in the digital world, *Internet Research: Electronic Networking Applications and Policy, 8* (2).

Lubbe S. (1999). Leverage effect, Working paper - Department of Computer Science and Information Systems, Vista University, Bloemfontein

Oxford University Press. (1996). Oxford Dictionary and Thesaurus.

Reich B.H. & Benbasat L. (1996). *Measuring the Linkage between Business and Information Technology Objectives,* MIS Quarterly 20(1)

Sohn C. (1998). How Information Systems provide competitive advantage: An Organizational learning perspective, *Proceedings of 1998 Americas Conference, Baltimore,* Maryland, August 14-16

Tallon P. T. & Kraemer K.L. (1998). A Process-oriented Assessment of the Alignment of Information Systems and Business Strategy: Implications for Business Value, *Proceedings of 1998 Americas Conference,* Baltimore, Maryland, August 14-16

Tallon P.P, Kraemer K.L. & Gurbaxani V. (1999). Fact or Fiction: The reality behind Executive Perceptions of IT Business Value, Research paper July 29, 1999

Ulrich & Lake (1990) *cited* by Sohn C. 1998 in How Information Systems provide competitive advantage: An Organizational learning perspective, *Proceedings*

Copyright © 2003, Idea Group Inc. Copying or distributing in print or electronic forms without written permission of Idea Group Inc. is prohibited.

*of 1998 Americas Conference,* Baltimore, Maryland, August 14- 16

Whiteley D. (1999) Merging Electronic Commerce Technologies for Competitive advantage, *Proceedings of 1998 Americas Conference,* Baltimore, Maryland, August 14-16

White G.K. & Manning B.J. (1998). Commercial WWW site appeal: How does it affect online food and drink consumers' purchasing behaviour? *Internet Research: Electronic Networking Applications and Policy, 8* (1).

Copyright © 2003, Idea Group Inc. Copying or distributing in print or electronic forms without written permission of Idea Group Inc. is prohibited.

## Chapter VI

# Online Customer Service

Rick Gibson
American University, USA

## ABSTRACT

*The purpose of this study is to find an effective online customer service strategy. Although the effectiveness of the online customer service will vary and depend on the type of business the company is involved in, the usage of different types of tools in this arena have proven to be more useful than others. Effectiveness in this work will be used, in the sense that the more effective strategy will lead to more satisfied customers, a higher customer retention rate and higher revenue for the business.*

Copyright © 2003, Idea Group Inc. Copying or distributing in print or electronic forms without written permission of Idea Group Inc. is prohibited.

# ONLINE CUSTOMER SERVICE

Why does online customer service have to be researched? Online customer service has become an integral part of success for companies conducting business on the Web. The Internet is changing the way firms interact with customers. However, according to the research that has been done on the subject, it is found that most companies are not fully utilizing the Internet fully for online customer service. For example, almost 70 percent of shoppers abandon their shopping carts before making a purchase, primarily due to the lack of online customer service (Bernett, 2000).

This finding is very engaging since it points out that many web businesses lose business because of the ineffectiveness of online customer service. Moreover, unlike in the off-line world, the customer can easily switch from one business to another with a click of a mouse (Motti, 2000). Another interesting fact is that more than 50 percent of e-commerce websites do not have any type of customer service that guides the customer to the web page where he/she can get help (Trott, 2000). In addition, it is estimated that $1.6 billion was lost in sales in 1999 due to the lack of online customer service (Lucent Technologies).

Web businesses have found out that customer service is as important in the virtual store as it is in the traditional "brick and mortar" store (Bernett, 2000). In addition, the purposes of providing quality customer service is the same as in the web as in the "brick and mortar" stores, which are attracting and retaining target customers in order to increase sales and profits (Boone & Kurtz, 1995). However, the methods of conducting customer service off-line and online differ greatly. The businesses that are not online usually use call centers where customer service agents assist the customers via telephone. On the contrary, online businesses rely on several methods in supporting their customers. An interesting study conducted by Forrester Research indicated that in 1997, 97 percent of customer interactions regarding customer service were via telephone. The remaining 3 percent were via e-mail (2%) and the Internet (1%). However, it is estimated that in 2003, 56 percent of customer interaction with the customer service department will be via the web. Thirty percent will be via e-mail; nine percent will be via cross-channels (with the usage of multiple means), and only five percent will be via telephone (Cincom).

Most traditional retailers view customer service as a problem solving action after the; sale. The businesses that want to assist their customers online effectively will have to disagree with the statement above. In e-commerce, customer service begins the moment that a visitor enters the website. The customers have to be assisted during shopping in case they have product questions and/or product locations. They have to be able to find help during the purchasing process in case

Copyright © 2003, Idea Group Inc. Copying or distributing in print or electronic forms without written permission of Idea Group Inc. is prohibited.

they have questions regarding the billing issues, receipts and the checkout process. Moreover, after ordering the products, they have to be able to learn about the status of order. In addition, after the customers receive the product, they have to be provided with the information regarding the product setup and product returns (Bernett, 2000).

Several tools are currently used in assisting customers online. These are: E-Mail, Frequently Asked Questions (FAQ), Co-Browsing (Escorted Browsing), Application Sharing, Text-Based Chat, Video-Conferencing, Voice Over the Internet Protocol (VOIP), and online customer service agents (Lucent Technologies).

E-mail is the most common and the easiest form of communication on the web at present. This method is usually preferred since the customer does not need a special hardware to send and receive e-mails. Web-based mail applications let the user to send and receive e-mails without using special software. . There are more advantages on the customer side with regard to using e-mail. First, the customer can send e-mail anytime. Another advantage is that it is much more convenient writing a quick e-mail than going through a telephone queue and being transferred among customer service agents.

Because of these factors of convenience, websites begin to receive thousands of e-mails everyday. However, dealing with a big load of e-mail has been problematic for most of the websites. A study by Jupiter Communications reviewed 125 leading web sites. It sent them relevant consumer questions by e-mail to see the response time of the websites. Forty-two percent of the sites never responded to the e-mail, took more than five days in responding or simply they did not have an e-mail address on their site for customer service purposes (Romeo, 1999). There are a number of technologies used to sort out the e-mails in order to achieve a higher response time and these will be discussed later in the study.

Frequently asked questions are the questions that are routinely asked by customers on a specific site. These questions can be handled in three possible ways. The first one is to place a frequently asked questions and answers area on the website so that a customer can look at that web page before contacting the customer service. The second way is to e-mail a preconfigured template document to the customer. The third way is to use an auto responder. These methods are effective, however it makes it hard to follow-up with the customers to see if they were satisfied with the service (Hunter, 1998).

Co-Browsing (Escorted Browsing), Application Sharing, Text-Based Chat, Voice Over the Internet Protocol (VOIP), Video-Conferencing, and online customer service agents all use live help in order to assist the customers. Co-

Copyright © 2003, Idea Group Inc. Copying or distributing in print or electronic forms without written permission of Idea Group Inc. is prohibited.

Browsing enables the agent and the customer to co-navigate an Internet site which means that they see the same web-page at the same time. This takes advantage of the push technology where the customer service representative can push pages to the customer (Lucent Technologies). This tool also allows the customer service representative to up-sell and cross-sell (Schefter, 2000).

Microsoft Net Meeting, and Netscape Collabra are the two leading off-the-shelf communication applications that are used in this technology. In addition, Java applets or ActiveX control can also be used by the website if the customer does not have a communication application that is capable of co-browsing (Genesys).

Application Sharing is different from co-browsing since in this case, an application can be viewed and controlled at the same time by the customer and the agent. For example, if an agent wants to show a drawing to the customer, they can both use a graphics application. The downside of application sharing is that it requires more bandwidth as well as an installed software in both the customer's and the agent's computer (Genesys).

Text-Based chat (text conferencing) is an older technology than application sharing where two or more people communicate by sending messages to each other. There are numerous advantages in this technology. The users do not have to have a PC with multimedia capabilities. In addition, firewall issues do not affect text-based chat. Some canned responses can be used by the agent to save time. Moreover, it can be used to supplement Voice Over IP applications where a word or a phrase that is used by one of the parties cannot be understood. In that case, the party who spoke can type what he/she said (Lucent, 2000). Chat technology can be used with different applications. It can be in an HTML form, a Java applet, a browser plug-in, or a stand-alone application such as ICQ or Mirc (Genesys).

Voice Over the Internet Protocol is one of the most useful tools that are used in the online customer service arena. This tool allows voice calls to be sent over the Internet. This is advantageous for the customer since he/she does not need to use an additional line to talk to the agent. It is beneficial for the agent because they can look up information while assisting the customer (Bennett, 2000). However, both parties have to have some hardware and software that allows Voice over IP. A VoIP-enabled application, a speaker and a microphone have to be installed in both parties' computers. In addition, the quality of the communication can be low depending on the traffic of the network and the voice compression. The 56k modems allow Voice Over IP. However, with such a modem, it is difficult to navigate the web and talk at the same time; one of the applications will suffer due to the lack of bandwidth. Most VOIP applications use the H.323 protocol. Microsoft Net meeting and Netscape Collabra are compliant with this protocol (Cincom).

Copyright © 2003, Idea Group Inc. Copying or distributing in print or electronic forms without written permission of Idea Group Inc. is prohibited.

Video conferencing over the net is uncommon at present in the customer service arena for a number of reasons. First, a video camera in both parties is necessary. Second, a quality video conferencing session requires a high bandwidth with powerful computers. Third, many Internet users may not want to show their faces to the agent. However, this can be solved by one-way conferencing where the customer can see the agent but the agent cannot see the customer. This approach eliminates the need for a video camera on the customer side. The protocol that supports Voice over IP (H.323) also supports videoconferencing. However, a H.323 compliant application such as Net Meeting is necessary to run a videoconferencing session (Genesys).

Online customer service agent is software that tries to assist the Internet customers in a timely and a more human way. This software (also called a bot) acts like a human customer service agent and tries to anticipate the customer's needs, based on their click trail on a website. Moreover, it is customizable and it uses facial expressions to add a human element to the interaction. Although this technology is in its infancy, it narrows down the problem for the human representative, thus cutting down the costs. The average call to a live customer service representative costs about $4, whereas an automated interaction costs about a quarter (Trott, 2000). Artificial intelligence plays a crucial role on the effectiveness of a bot. The most advanced bots are capable of learning as they interact. In addition, these bots also can remember the actions users took and the preferences that they made. It is also suggested that these bots should have scripts ready for predictable questions asked in their websites. This can be done by the help of a database. However, the same question can be asked in many different ways and this should not be overlooked (Zhivago, 2000)

Some tools that enable the interaction between the customer and the agent are front-end tools. There are other tools that are used by the agents to assist their customers better and that are invisible to the customers. The tools of "Customer Relationship Management" are essential to an effective web-based customer service. Customer Relationship Management is defined as'' a strategy reflecting the business' processes and technology that can be combined to optimize revenue, profitability, and customer loyalty (Microsoft). CRM is about creating relationships with customers on an individual basis in order to transform the information collected into a useful tool to treat each customer differently. This customer-focused approach is a key element in creating an effective online customer service. This approach creates customer loyalty, which. is extremely important since it is much harder to gain new customers than retaining the existing ones. Some statistics indicate the importance of customer loyalty. The top 20 percent of the customers

Copyright © 2003, Idea Group Inc. Copying or distributing in print or electronic forms without written permission of Idea Group Inc. is prohibited.

delivers 80 percent of the revenues. This is called the "80/20 Pareto Principle." In addition, the top 20 percent of the customers deliver most of the profits. Moreover, the existing customers deliver up to 90 percent of revenues. All of these figures show that customer satisfaction and loyalty should be of utmost importance to every business (Curry, 2000).

The database technology enables the online customer agents to acquire information about their customers in order to provide them a better and personalized service. Any type of information gathering whether it is from a simple shopping list for a collection of customer information is considered a customer database as long as it is computerized. This technology stores the information about the customers such as their transactions, interactions, profiles and purchase histories in order to provide them personalized customer service on the net. Data Warehousing and Data Mining are the two major technologies that are used in Customer Relationship Management that enable the online customer service agents to provide a better service to their customers (Coronel, 2000).

Data Warehouses were developed as data storage facilities. These data warehouses receive data from the operational databases as well as other sources in order to provide a comprehensive data pool from which information can be selected. The most crucial aspect of data warehouses is that the data that is stored in structures is able to generate specific information (Gardner, 1998).

Data mining is the process where the marketer tries to find trends and patterns in huge amounts of data. The goal of this process is to search through large quantities of data and discover new information that can be useful for the company. "The benefit of data mining is to turn this newfound knowledge into actionable results, such as increasing a customer's likelihood to buy" (Groth, 2000). Data mining also enables the company access to detailed knowledge of customer profiles, and the ability to analyze customer satisfaction.

There are also some essential components of online customer service that have to be in place regardless of the business. The website has to provide simple navigation, fast image download times, and fast access to information. The human aspect should also be of utmost importance such as quick e-mail responses, and toll-free phone numbers for customer service. Furthermore, the product information has to be detailed on the website, such as the descriptions, specifications, pricing, FAQs, smart shopping carts, and cross-purchasing suggestions. Another crucial component is the issue of incentives such as gift-wrapping, promotions, money-back and on-time delivery guarantees, and offers of free shipping and handling. Moreover, a trust has to be built between the site and the customer. This can be achieved by placing the explanations regarding the secure transactions, and

Copyright © 2003, Idea Group Inc. Copying or distributing in print or electronic forms without written permission of Idea Group Inc. is prohibited.

the information regarding privacy of data collected. Lastly, a follow-up mechanism has to be in place at the website such as confirming the e-mails, an automated shipping/tracking system, prompt delivery, and e-mail satisfaction surveys (LeClaire & Picozzi, 2000).

As it can be seen, the online customer service arena is a very important component of achieving customer satisfaction and retaining customers. Many tools and strategies can be used to achieve an effective online customer service. The question is how effective are these tools and where they have to be used. Each tool has advantages and disadvantages. An online business has to evaluate the pros and the cons of each tool and then decide the most appropriate tools for its website.

# REFERENCES

Bernett, H. (2000). E-Commerce, Customer Service, and the Web-Enabled Call Center. Presentation at American University.

Boone, L. E. & Kurtz, D. L. (1995). *Contemporary Marketing Plus*. Forth Worth, TX: The Dryden Press. .

Cincom. Transforming Your Call Center Into a Multimedia Contact Center. Available at: http://www.crmxchange.com/whitecacers/cincom3.html.

Coronel, R. (2000). *Database Systems: Design. Implementation and Management*. Cambridge, MA: Course Technology.

Curry, J. (2000). *The Customer Marketing Method: How to Implement and Profit from Customer Relationship Management*. New York, NY: The Free Press.

Gardner, S. (1999). E Building the data warehouse. *Communications of the ACM. 41* (9), 52-60.

Genesys. Customer Service on the Internet; Redefining The Call Center. Available at: http://www.telemkt.com/whitepapers/genesys-redefining.html.

Groth, R. *Data Mining: Building Competitive Advantage*. Saddle River, NJ: Prentice Hall.

Hunter, S. (1999). Improving Your Online Customer Service: Seven Common Mistakes Businesses Make With E-Mail That Can Cost You Sales. Available at: [On-Line].

LeClaire, J. & Pirozzi, L. (2000). Understand Online Customer Service. Available at: http://w-vvw.workz.com/html/404.html.

Lucent Technologies. (2002). Web-Enabled Customer Contact Centers. Available at: http://www.crmxchange.com/whitepapers/lucent.htmi.

Microsoft Corporation (Year). Overview of Customer Relationship Management.

Copyright © 2003, Idea Group Inc. Copying or distributing in print or electronic forms without written permission of Idea Group Inc. is prohibited.

Available at:http://www.microsoft.com/business/crm/resources/overview.asp.
Mottl, J. (2000). Customer Service Is E-Tailer's Costly Challenge. Available at:.
    http://www.intemetweek.com.
Romeo, J. (1999). Online customer service comes of age. Available at: http://
    www.ecommercetimes.com/printer.
Schefter, P. (2000). E-loyalty. your secret weapon on the web. *Harvard Business
    Review*, 105.
Trott, B. (2000). Online 'agents' evolve for customer service. *Infoworld*. 31.
Zhivago, K. (2000). To bot or not to bot? *Business* 2.0. 54.

Copyright © 2003, Idea Group Inc. Copying or distributing in print or electronic forms without written permission of Idea Group Inc. is prohibited.

Chapter VII

# E-Commerce and Executive Information Systems: A Managerial Perspective

Geoff Erwin
Cape Technikon, South Africa

Udo Averweg
University of Natal, South Africa

## ABSTRACT

*The rapid spread of connectivity via the World Wide Web has dramatically altered the ways in which organizations deal with customers and the methods that executives adopt to be informed about business operations. This chapter reviews Executive Information Systems (EIS) and the way in which EIS interacts with e-commerce applications.*

## INTRODUCTION

The emergence of e-commerce as a business transactions mechanism has introduced new considerations into strategic decision-making and the design of decision support information systems. Information Systems (IS) at strategic decision-making levels are commonly characterized as being 'executive'. The impact of such decisions, by definition, will be significant (Turban, 2001). IS for

Copyright © 2003, Idea Group Inc. Copying or distributing in print or electronic forms without written permission of Idea Group Inc. is prohibited.

executives therefore require careful planning in their features and applicability to organizational situations. Executive Information Systems (EIS) are designed to serve the needs of executive users in strategic planning and decision-making (Srivihok, 1998) and for making strategic and tactical decisions (Salmeron et al., 2001).

The technology for EIS is evolving rapidly and future systems are likely to be different (Sprague & Watson, 1996). EIS is now clearly in a state of flux. Turban (2001) concurs: "EIS is going through a major change". Emerging (Web-based) technologies can redefine the utility, desirability and economic viability of EIS technology (Volonino et al., 1995). These technologies may need to be applied to resolve one of the most challenging and critical components of an EIS - data accessibility - which often incurs the greatest time, expense and delay in EIS implementation. The accessibility, navigation and management of data and information for improved executive decision-making are becoming *critical* in the new global business environment (Averweg & Erwin, 2000).

The evolution of Web-based business activity has resulted in the term *e-Business* being used to refer to three categories of business activity:

- Business-to-Employee (B2E): Intranet-based applications internal to a company;
- Business-to-Consumer (B2C): Internet-based applications for a company's customers; and
- Business-to-Business (B2B): Extranet-based applications for a company's business partners. (An IOS: Inter Organizational System).

The term *e-commerce* often refers only to B2B (McNurlin & Sprague, 2002). However, in this Chapter the term e-commerce is used to refer to all three categories. EIS, specifically in the e-commerce environment, is the focus of this chapter.

# BACKGROUND

Concern about Information Technology (IT) impact on society, organizations and people is not new. Some 170 years ago, British intellectuals expressed philosophical arguments about the effects of the Industrial Revolution on society (Turban et al., 1999). While there are philosophical, technological, social and other differences between that society and our own, there are people nowadays who believe that mankind is threatened by the evolution of technology. However, our society has not rejected technology and we recognize that computers and technology are essential to maintaining and supporting our culture. IT has become the major

Copyright © 2003, Idea Group Inc. Copying or distributing in print or electronic forms without written permission of Idea Group Inc. is prohibited.

facilitator of business activities in the world today. It is assumed that members of an organization will reap the fruits of new technologies and that computers have no negative impact on society, organizations and people.

IS are technology-based innovations created and used by individuals, organizations and societies (Allen, 2000). Technological advances have allowed organizations to utilize IT applications in all aspects of organizational management (Khosrowpour, 1998). IS to support senior executives have been available for well over a decade (Poon & Wagner, 2001). EIS grew out of the development of IS to be used directly by executives and used to augment the supply of information by subordinates (Srivihok, 1998). EIS is a technology that is continually emerging in response to managers' specific decision-making needs (Turban et al., 1999).

The ubiquitous nature of the Internet and its universal connectivity/networked capability are dramatically revolutionizing the manner in which organizations and individuals access and share information (Anandarajan & Simmers, 2001). Many organizations hope to achieve general distributed computing across large networks or even across organizational boundaries (PriceWaterhouseCoopers, 2000). The adoption of networks in day-to-day and critical organizational operations has made them indispensable for co-operative work (Theoharakis & Serpanos, 2002).

Advances in computer technologies combined with telecommunication technologies have lead to the development of the Internet and its most popular application, the World Wide Web ('the Web') (Khosrowpour, 2000). A new set of challenges has arizen as organizations integrate IT into all functions and activities of the modern organization (Khosrowpour & Liebowitz, 1997).

As the usage of IT increases, Web-enabled information technologies can provide the means for greater access to information from disparate computer applications and other information resources (Eder, 2000). Many organizations have benefited from the technologies of the Web (Khosrowpour, 1998). These technologies include: intranet, Internet, extranet, e-commerce, Wireless Application Protocol (WAP) and other mobile technologies. E-commerce and its impact on EIS implementation and usage is the focus of this chapter. There exists a high degree of similarity between the characteristics of a 'good EIS' and Web-based technologies (Tang et al., 1997).

Executives deal mostly with ill-structured decision-making (Lee & Chen, 1997). Definitions of e-commerce and e-business are varied (see, for example, Rockart & DeLong, 1998; Carlsson & Widmeyer, 1990; Watson et al., 1991; Westland & Walls, 1991; Whymark, 1991; Millet & Mawhinney, 1992; Rainer et al., 1992). All definitions identify the need for information that supports decisions about the business as the most important reason for the existence of EIS (Khan, 1996). Executives use EIS to extract, filter, compress and track critical data (Butler,

Copyright © 2003, Idea Group Inc. Copying or distributing in print or electronic forms without written permission of Idea Group Inc. is prohibited.

1992). EIS also allow executives seamless access to complex multi-dimensional models so that they can see their business at a glance (Harris, 2000). EIS applications support executive information needs and decision-making activities (Gillan & McPherson, 1993). An effective way to evaluate the success of an EIS is to obtain opinions from the executive users (Monash University, 1996). In this chapter, EIS is defined as "a computerized system that provides executives with... access to internal and external information that is relevant to their critical success factors" (Watson et al., 1997). While a definition is useful, describing the capabilities and characteristics of EIS provides a richer understanding.

Earlier studies described EIS capabilities that are focused on providing information that serves executive needs. Srivihok (1998) reports that these capabilities are concerned with both the quality of the system (e.g., user friendliness) and information quality (e.g., relevance). Sprague and Watson (1996) identify the following capabilities or characteristics of EIS:
• tailored to individual executive users;
• extract, filter, compress and track critical data;
• provide online status access, trend analysis, exception reporting and 'drill down';
• access and integrate a broad range of internal and external data;
• user-friendly and require little or no training to use;
• used directly by executives without intermediaries; and
• present graphical, tabular and/or textual information.

Other researchers suggest additional capabilities and characteristics of EIS:
• flexible and adaptable (Carlsson & Widmeyer, 1990);
• should contain tactical or strategic information that executives do not currently receive (Burkan, 1991);
• facilitate executives' activities in management such as scanning, communication and delegating (Westland & Walls, 1991);
• make executive work more effective and efficient (Friend, 1992);
• assist upper management to make more effective decisions (Warmouth & Yen, 1992);
• incorporate an historical 'data cube' and 'soft' information (Mallach, 1994) ('datacube' is a structure in which data is organized at the core of a multi-dimensional online analytical processing (OLAP) system (Ross, 2001) and 'soft' information includes opinions, ideas, predictions, attitudes, and plans);
• provide support for electronic communications (Rainer & Watson, 1995a); and

Copyright © 2003, Idea Group Inc. Copying or distributing in print or electronic forms without written permission of Idea Group Inc. is prohibited.

•    enhanced relational and multi-dimensional analysis and presentation, friendly data access, user-friendly graphical interfaces, imaging, hypertext, intranet access, Internet access and modelling (Turban et al., 1999).

The terms 'Executive Information Systems' and 'Executive Support Systems (ESS)' are sometimes used interchangeably (Turban et al., 1999). However, ESS usually refers to a system with a more extensive set of capabilities than an EIS (Mallach, 1994). Rainer and Watson (1995a) and Watson et al. (1991) report that these capabilities include the provision of data analysis capabilities (*eg.*, spreadsheets, query languages and DSS) and the provision of organizational tools (*eg.*, electronic calendars, personal information filing and management) (Beheshti, 1995). Kuo (1998) proposes an ecological model of managerial intuition for the purpose of developing effective ESS. Turban et al. (1999) define ESS as 'a comprehensive support system that goes beyond EIS to include analysis support, communications, office automation, and intelligence'.

Other titles for EIS are 'Everyone's Information System' (Wheeler et al., 1993), 'everyone information system' (Frolick & Robichaux, 1995), 'enterprise-wide' EIS (Frolick & Robichaux, 1995) and 'Enterprise Information Systems' (Post & Anderson, 1997; O'Brien, 1999). O'Brien (1999) states that (the evolution of various) names reflect the fact that more features, such as Web browsing, electronic mail and groupware tools, are being added to many systems to make them more useful to executive managers. Turban et al. (1999) report that sometimes where EIS applications embrace a range of products targeted to support professional decision-makers throughout the organization, the term 'Everybody's Information System' is used. For the purposes of this chapter, the acronym EIS shall mean "Executive Information Systems."

## ISSUES FOR E-COMMERCE AND EIS

Clearly, the Internet offers a wealth of new opportunities for e-commerce, but also presents executives with a series of new challenges (Laudon & Laudon, 2000). These challenges stem from the fact that Internet technology and its business functions are relatively new.

It is commonly argued that IT affects the organization structure and strategy and has a profound effect on management (Drucker, 1988; Leavitt & Whistler, 1988). It alters the nature of work in the workplace (Igbaria et al., 1994). The Internet has fundamentally changed the ways in which much of the world communicates and does business (Hardaker, 2001). EIS can help managers keep information of relevance to the organization's success within easy access (Beheshti,

Copyright © 2003, Idea Group Inc. Copying or distributing in print or electronic forms without written permission of Idea Group Inc. is prohibited.

1995). For organizations to succeed in the current business environment, it is important to develop policies to create an environment conducive to technological innovation (Basu et al., 2000). This entails the need to prioritize available resources to facilitate the development of implementation of new ideas and technologies (Lai et al., 1993). The continued challenge remains in terms of ensuring adaptability and flexibility of information interfaces - both internally and externally - required for coping with dynamically changing business and competitive environments (Malhotra, 2001). As Mintzberg (1973) noted three decades ago, a manager performs several roles in which the exchange and access to information is a critical aspect. Executives (as most people) nowadays have an increased computing literacy and can confidently locate online information due to extensive personal use of the Web - a knowledge that is transferred to their business setting (Basu et al., 2000).

In research by Basu et al. (2000), traditional EIS saw limited diffusion within the organization from the late 1980s to early or mid-1990s. In some instances, only a few executives actually used the system. In other instances, more users were logged but only on a small percentage of the system's functions. The primary reasons cited were that the system was too hard to use or had little added value for the executive. Actual benefits were not aligned with high development costs.

During this period, globalization was increasing the expanse of business along with the need for more collaboration and communication. Concurrent with these conditions, Internet technologies began to emerge. Once the infrastructure for Internet access existed, it was a small step to apply the technology to internal applications. Nowadays, executives can confidently locate online information due to extensive personal use of the Web. Furthermore vendors of decision support software see a larger market for products that provide OLAP, spreadsheet-type capabilities and Web-based solutions (Gray & Watson, 1998).

With the demise of early EIS, in the early to mid-1990s some organizations surveyed (Basu et al., 2000) consciously discontinued these systems after conducting some level of value-based analysis. Other organizations gradually migrated to intranet applications such as groupware, database querying, electronic calendars and scheduling. All organizations have browser-based intranets. The Web browser has become 'an almost universal interface' for end-user access (Gray & Watson, 1998). In summary, the catalysts of change away from traditional EIS implementations were as follows:

- systems difficult to use;
- high costs compared to value added;
- lack of needed information;
- dissatisfied executives;

Copyright © 2003, Idea Group Inc. Copying or distributing in print or electronic forms without written permission of Idea Group Inc. is prohibited.

- globalization; and
- better technological alternatives were becoming universally known.

Some studies suggest that others should access EIS besides executive users (see Hasan, 1995; Volonino et al., 1995; Rai & Bajwa, 1997). These researchers suggest that EIS should be viewed as technology to be used to solve major business problems arising from global competitive and recessionary forces. Salmeron (2001) notes EIS as the technology for information delivery for all business end users. Kennedy (1995), and Messina and Sanjay (1995) report that EIS have spread throughout organizations. The non-executive users or data providers include personnel from functional areas that include treasurers, accounting managers and controllers. It is evident that EIS requires continuous input from three different stakeholder groups (known as constituencies):

- EIS executives and business end-users;
- EIS providers (*ie.* persons responsible for developing and maintaining the EIS); and
- EIS vendors or consultants.

This multiple constituency approach has been used to investigate other types of IS (see, for example, Hamilton & Chervany, 1981; Alavi, 1982; Hogue & Watson, 1985; Watson et al., 1987). Rainer and Watson (1995b) suggest that these three EIS stakeholder groups may have different ideas on factors affecting successful EIS development and operation.

EIS flexibility should be considered in the development of an EIS in an organization (Srivihok, 1998). Salmeron (2001) reports that if this were not so, EIS would soon become a useless tool that would only deal with outdated problems and would therefore not contribute to decision-making. EIS should be flexible to support different classes of business data (e.g., external, internal, structured and unstructured) and different levels of users (e.g., executives and non-executive users). Turban et al. (1999) report that two types of EIS can be distinguished:

(1) The one designed especially to support top executives; and
(2) The EIS that is intended to serve a wider community of users.

Realizing the benefits from EIS, these users become the driving force behind EIS development and implementation (Messina & Sanjay, 1995). With the emergence of global IT, existing paradigms are being altered which are spawning new considerations for successful IT implementation (Averweg & Erwin, 2000). Web-based technologies, especially e-commerce, are causing a revisit to existing IT implementation models, including EIS. The Web is 'a perfect medium' for

Copyright © 2003, Idea Group Inc. Copying or distributing in print or electronic forms without written permission of Idea Group Inc. is prohibited.

deploying decision support and EIS capabilities on a global basis (Turban et al., 1999).

IT is a catalyst for fundamental changes in the structure, operations and management of organizations (Dertouzos, 1997). Wreden (1997) reports that IT capabilities support five business objectives: improving productivity (in 51% of organizations), reducing cost (39%), improving decision-making (36%), enhancing customer relationships (33%) and developing new strategic applications (33%). Elliott (1992) notes that EIS use enables more analysis in decision-making, faster problem identification and faster decision-making. Executives who use EIS most frequently should notice the greatest increase in their decision-making speed (Leidner & Elam, 1993). Leidner and Elam (1993) report that frequency of EIS use and length of time of use are both significantly associated with an executive's decision-making process.

With the evolution of distributed computer technology, paved by the rapid adoption of Web technology, there is a growing need for improved decision-making at any time, anywhere and with any participants. Digital communication requires a new business paradigm: being able to join and use any type of business system any time, anywhere (Worthington-Smith, 2000). This is a real business problem and is especially relevant to EIS. Palvia et al. (1996) argue that organizations need to incorporate a global dimension in the design of their EIS. These researchers suggest that such global EIS will incorporate international information that will be critical to executives of multinational and global organizations in order for them to conduct business and compete globally.

E-commerce is changing the manner in which business is done, especially when facilitated by appropriate support systems (Turban & Aronson, 1998). Turban and Aronson (1998) note that the decision support of groups whose members are at different locations has become an increasingly important topic because of the increased emphasis on workgroups and teams. The computing environment is rapidly changing to a global network and the Internet (World Wide Web) is the enabling link of accessing information anywhere and at any time.

There is an increasing use of teams and teamwork in organizations (see, for example, Navarro, 1992; Roberts, 1995). The major feature of networked decision support is the use of several computers and databases connected by networks (Turban & Aronson, 1998). When people are working in teams (especially when the members are in different locations and may be working at different times), they need to communicate, collaborate and access a diverse set of information sources in multiple forms (e.g., text, video, graphics and voice). Turban and Aronson (1998) indicate that the major information architecture that supports this distributed decision support environment consists of the Internet and intranets.

Copyright © 2003, Idea Group Inc. Copying or distributing in print or electronic forms without written permission of Idea Group Inc. is prohibited.

The Internet is used to support interorganizational decision-making and provides access to information outside the organization. Intraorganizational networked decision support is achieved by the use of an intranet that allows people within an organization to work with Internet tools and procedures. Many software and hardware computing tools can be used to provide networked decision support. These software tools are often known as *groupware*.

Networked computing allows organizations to conduct business electronically among business partners. E-commerce is the execution of business via computer networks. E-commerce means more than just another way to reach the market since it also gives businesses the ability to share and manage knowledge, the most valuable business commodity (Emery, 2001).

The Internet, intranets and extranets provide many benefits. The Internet's global connectivity, ease of use, low cost and multimedia capabilities can be used to create interactive applications, products and services (Laudon & Laudon, 2000).

There is a popular misconception that e-commerce is merely having a Web site. However, this is not the case. There are many e-commerce applications (e.g., banking, shopping in online stores, malls, finding a job, conducting an auction, collaborating electronically on research and development projects) and in order to execute these applications, it is necessary to have supporting information, organizational infrastructure and systems. Applications of e-commerce may be divided into three categories: (1) buying and selling goods and services. These are commonly referred to as electronic markets; (2) facilitating inter- and intraorganizational flow of information, communication and collaboration. These are sometimes referred to as interorganizational systems (IOS); and (3) providing customer service. Attention is now focused on IOS.

An IOS involves information flow between two or more organizations (Turban et al., 2000). Its major objective is efficient transaction processing (e.g., transmitting orders, invoices and payments) using electronic data interchange (EDI) or extranets (Senn, 1996). An IOS is a unified system encompassing several business partners. A typical IOS will include an organization and its suppliers and/or customers. The most prominent types of IOS are:

- Electronic data interchange (EDI), which provides a secured B2B connection over value-added networks (VANs). Before the introduction of the Internet, e-commerce took the form of EDI over private networks (Mougayar, 1998);
- Extranets which provide secured B2B connection over the Internet;
- Electronic funds transfer (EFT);
- Integrated messaging - delivery of e-mail and fax documents through a single electronic transmission system that can combine EDI, electronic mail and

Copyright © 2003, Idea Group Inc. Copying or distributing in print or electronic forms without written permission of Idea Group Inc. is prohibited.

electronic forms;

- Shared databases - information stored in repositories is shared between trading partners and is accessible to all parties. The sharing is primarily done over extranets; and
- Supply chain management (SCM) - co-operation between an organization and its suppliers and/or customers regarding demand forecasting, inventory management and orders fulfillment can reduce inventories, speed shipments and enable just-in-time manufacturing.

Better information leads to better decisions. Information from outside the organization should be included in EIS (Chiusolo & Kleiner, 1995). Chiusolo and Kleiner (1995) note the need for current and accurate information about the organization and the world in general is common among all executives.

The global nature of e-commerce technology, its interactivity, resourcefulness and rapid growth of supporting infrastructures (especially the Web) result in many potential benefits to organizations (Turban et al., 2000) (e.g., expanding the marketplace and reducing inventories). There are, however, limitations of e-commerce. These can be grouped into technical and non-technical categories. Some of the technical limitations of e-commerce for EIS are:

- Insufficient telecommunication bandwidth;
- Limited search facilities for filtering Web information;
- Lack of system, reliability, standards and some communication protocols;
- Continued evolution and rapid change of software development tools; and
- Integration difficulty between the Internet and e-commerce software and some existing applications and databases.

## FUTURE TRENDS

E-commerce could become a significant global economic element during the twenty-first century (Clinton & Gore, 1997). The infrastructure for e-commerce is *networked computing* which is emerging as the standard computing environment in business, government and home (Turban et al., 2000). Networked computing connects several computers and other electronic devices by telecommunications networks. This facilitates users accessing information stored in several places and to communicate and collaborate with others. This new breed of computing is helping many organizations (private and public) not only to excel but also frequently to survive. Since the technologies required for networked computing are often immature, organizations are only beginning to develop systems that are large steps toward achieving an ideal that can probably only be approximated

Copyright © 2003, Idea Group Inc. Copying or distributing in print or electronic forms without written permission of Idea Group Inc. is prohibited.

(PriceWaterhouseCoopers, 2000). Clearly this new networked computing environment has impacted existing IS and in turn impacts an organization's EIS.

Turban et al. (2000) predict that e-commerce limitations will lessen or be overcome. Appropriate planning can minimize their impact. These researchers report that rapid progress in e-commerce is taking place. As technology improves and experience accumulates, the ratio of e-commerce benefits to costs will increase resulting in a greater rate of e-commerce adoption. [For a discussion of e-commerce in developing countries, see Worthington-Smith (2001).]

Watson et al. (1997) suggest the following trends in EIS:

- EIS are becoming more enterprise-wide with greater decision support capabilities;
- EIS are becoming used or integrated with software not specifically designed for it, e.g., the World Wide Web technology; and
- EIS are gaining in intelligence through the use of intelligent software agents.

Executives place substantial requirements on EIS (Turban & Aronson, 1998). First, they often ask questions which require complex, real-time analysis for their answers. Hence, many EIS are being linked to data warehouses and are built using real time OLAP in separate multi-dimensional databases along with organizational DSS. There are also efforts to use data warehouse and OLAP engines to perform data mining (Han, 1998). Secondly, executives require systems that are easy to use, easy to learn and easy to navigate. Turban and Aronson (1998) report that current EIS generally possess these qualities. Thirdly, executives tend to have highly individual work styles. While the current generation of EIS can be moulded to the needs of the executive, it is difficult to alter the look and feel of the system or to alter the way in which the user interacts with the system. Fourthly, any IS is essentially a social system. The researchers note that one of the key elements of an EIS is the electronic mail capability it provides for members of the executive team. Nowadays, the electronic mailing of multimedia documents is becoming critical. Given this scenario, the EIS of the future will look significantly different from today's systems.

As organizations become more global in nature, providing information about international locations around the world is becoming critical to organizations' success. The accuracy and timeliness of information for decision-making become critical. The challenge has become to find ways to integrate information across the enterprize (Ba et al., 1997). The transparency of the integration of the information process is what makes Web technology so effective. Palvia et al. (1996) investigated the types of data that executives require in two scenarios: (1) introducing a new service or product into other countries; and (2) distribution channel expansion into other countries. Most of the executive information requirements include

Copyright © 2003, Idea Group Inc. Copying or distributing in print or electronic forms without written permission of Idea Group Inc. is prohibited.

demographic and marketing data from public sources and 'soft' information from personal contacts. Palvia et al. (1996) indicate that EIS can be used to provide the 'soft' information.

Basu et al. (2000) note that traditional EIS are 'a thing of the past' and most organizations have moved to other types of systems, such as WIIS. However, Salmeron (2001) reports 'a trend towards direct use of EIS by Spanish executives'. Lederer et al. (1998) report that organizations adopting WIIS indicate that the most important benefit of being on the Web is to 'enhance competitiveness or create strategic advantage'. Given the Web's ease of use in multiple environments and positive user perceptions, Basu et al. (2000) suggest that WIIS will be more widely adopted and diffused among organizations. They caution, however, that whether WIIS actually meet the information requirements of executives will de determined over the next five years as more organizations implement and experience these systems. Users will continue to pull and receive pushed information from the Web (Kendall & Kendall, 1999). The research by Basu et al. (2000) clearly shows that the advent of the Internet Age and resultant Web-based technologies have significantly impacted traditional EIS in organizations in the USA. Whether such a similar EIS situation exists in organizations in South Africa is currently the focus of a research project.

# CONCLUSION

It is important to recognize the role of the Web in decision support (Gray and Watson, 1998). Web-based technologies affect how all applications are developed and used. Gray and Watson (1998) note that many organizations are building decision support applications that have a multi-tier architecture consisting of a browser, Webserver and database. Recognizing this shift, vendors of decision support software (e.g., EIS) are making their products Web-enabled. Applications can now be accessed by browsers but still provide the capabilities long associated with decision support software (Gray & Watson, 1998).

Nowadays there is a need for organizations to adapt to constantly changing business conditions. IT can enable the fast adaptation necessary to accommodate these constant and rapid changes in the business environment (Keen, 1991). It is argued by the authors that EIS, particularly those that incorporate emerging Web-based technologies such as e-commerce, can increase an organization's ability to react to changing circumstances, quickly and flexibly. Organizations are evolving into virtual enterprises using integrated computer and communications technologies and linking hundreds, thousands, even tens of thousands of people together (Bleecker, 1994). Executives are now able to process queries that provide up to

Copyright © 2003, Idea Group Inc. Copying or distributing in print or electronic forms without written permission of Idea Group Inc. is prohibited.

minute information in situations where timeliness and completeness are of equal value (Volonino et al., 1995). Clearly the 1990s 'face' of EIS has changed and the next generation of EIS is on the doorstep of the e-commerce environment.

# REFERENCES

Alavi, M. (1982). An assessment of the concept of decision support systems as viewed by senior level executives. *MIS Quarterly*, *6* (4), 1-8.

Allen, J.P. (2000). Information systems as technological innovation. *Technology & People*, *13* (3), 210-221.

Anandarajan, M. & Simmers, C.A., (2001). *Managing Web Usage in the Workplace: A Social, Ethical and Legal Perspective*. Hershey, PA: Idea Group Publishing.

Averweg, U.R.F. & Erwin, G.J., (2000). Executive Information Systems In South Africa: A Research Synthesis For The Future. *Proceedings of the South African Institute of Computer Scientists and Information Technologists Conference (SAICSIT-2000)*, Cape Town, South Africa, 1-3 November.

Ba, S., Lang, K.R. & Whinston, A.B. (1997). Enterprise decision support using Intranet technology. *Decision Support Systems*, *20* (2), 99-134.

Basu, C., Poindexter, S., Drosen, J. & Addo, T. (2000). Diffusion of executive information systems in organisations and the shift to Web technologies. *Industrial Management & Data Systems*, *100* (6), 271-276.

Beheshti, H.M. (1995). Downsizing with executive information systems. *Industrial Management & Data Systems*, *95* (5), 18-22.

Bleecker, S.E. (1994). The Virtual Organization. *Futurist*, *28* (2), 9-14, March/April.

Burkan, W.C. (1991). *Executive Information Systems From Proposal Through Implementation*, Van Nostrand Reinhold.

Butler, J. (1992). Executive essay. *Mortgage Banking*, 72-74.

Carlsson, S.A. & Widmeyer, G.R. (1990). Towards a theory of executive information systems. *Proceedings of the Twenty-third Annual Hawaii International Conference on System Sciences (HICSS-23)*, Hawaii, USA, 195-201.

Chiusolo, E. & Kleiner, B.H. (1995). The most useful software for executives. *Industrial Management & Data Systems*, *95* (10), 25-28.

Clinton, W.J. & Gore, A., Jr., (1997). *A Framework for Global Electronic Commerce*. Available at: http://www.iitf.nist.gov/eleccomm/ecomm.htm.

Copyright © 2003, Idea Group Inc. Copying or distributing in print or electronic forms without written permission of Idea Group Inc. is prohibited.

Dertouzos, M. (1997). *What Will Be: How the New World of Information Will Change Our Lives*. San Francisco, USA: Harper Edge.

Drucker, P.F. (1988). The coming of the new organization. *Harvard Business Review*, *66* (1), 45-53.

Eder, L.B. (2000). *Managing Healthcare Information Systems with Web-Enabled Technologies*. Hershey, PA: Idea Group Publishing.

Elliott, D.G. (1992). Executive information systems: their impact on executive decision making. Unpublished thesis, The University of Texas at Austin, USA.

Emery, S. (2001). Trade World. Extending the boundaries for local business. *Global Trade, Executive Vision for International Business*, Production Unique, Cape Town, South Africa, 81-82.

Friend, D. (1992). Building an EIS your CFO will really use. *Chief Information Officer Journal*, *19* (1), 32-36.

Frolick, M.N. & Robichaux, B.P. (1995). EIS information requirements determination: Using a group support system to enhance the strategic business objectives method. *Decision Support Systems*, *14* (2), 157-170.

Gillan, C. & McPherson, K. (1993). Getting a clear picture of your business. *Computer Mail*, November.

Gray, P. & Watson, H.J. (1998). *Decision Support in the Data Warehouse*. V, New Jersey: Prentice-Hall.

Hamilton, S. & Chervany, N.L. (1981). Evaluating Information System Effectiveness-Part II: Comparing Evaluator Viewpoints. *MIS Quarterly*, *5* (4), 79-86.

Han, J. (1998). Towards On-Line Analytical Mining in Large Databases. *SIGMOD Record*, *27* (1), 97-107.

Hardaker, M. (2001). Dot Bomb. *SA Computer Magazine*, *9* (3), 48-52, April.

Harris, L. (ed.) (2000). Rand Water derives Bottom-Line benefits from EIS. *E-Business or Out of Business*, an Advertising Supplement compiled and published by Computing S.A., TML Business Publishing, Pinegowrie, South Africa, 28-29.

Hasan, H. (1995). *Organizational* issues and implementations of EIS. *Proceedings of the Sixth Australasian Conference on Information Systems*, Perth, Australia, *1*, 207-217.

Hogue, J.T. & Watson, H.J. (1985). Management's Role in the Approval and Administration of Decision Support Systems. *Information & Management*, *8* (4), 205-212.

Igbaria, M., Parasuraman, S. & Badawy, M.K. (1994). Work experiences, job involvement and quality of work life among information systems personnel. *MIS Quarterly*, *18* (2), xi-xii.

Copyright © 2003, Idea Group Inc. Copying or distributing in print or electronic forms without written permission of Idea Group Inc. is prohibited.

Keen, P.G.W. (1991). *Shaping the Future*. Boston, MA: Harvard Business School Press.

Kendall, J.E. & Kendall, K.E. (1999). Information delivery systems: An exploration of web pull and push technologies. *Communications of the AIS, 1* (14), 1-43.

Kennedy, D.H. (1995). Everybody's information systems? *Management Accounting London, 73* (5), 4.

Khan, S.J. (1996). The Benefits and Capabilities of Executive Information Systems. MBA *dissertation*, University of Witwatersrand, Johannesburg, South Africa.

Khosrowpour, M. (1998). *Effective Utilisation and Management of Emerging Information Technologies*. Hershey, PA: Idea Group Publishing.

Khosrowpour, M. (2000). Web-enabled technologies assessment and management: Critical issues. *Managing Web-Enabled Technologies in Organizations: A Global Perspective*, Hershey, PA: Idea Group Publishing.

Khosrowpour, M. & Liebowitz, J. (1997). *Cases on Information Technology In Modern Organizations*. Hershey, PA: Idea Group Publishing.

Kuo, F.Y. (1998). Managerial intuition and the development of executive support systems. Decision Support Systems, *24* (2), 89-103.

Lai, V., Guynes, J.L. & Bardoloi, B. (1993). ISDN: Adoption and diffusion issues. *Information Systems Management, 10* (4), 46-52.

Laudon, K.C. & Laudon, J.P. (2000). *Management Information Systems*. 6th edition, Upper Saddle River, New Jersey: Prentice-Hall.

Leavitt, H. H. & Whistler, T. I. (1988). Management in the 1980s. *Harvard Business Review, 36* (6), 41-48.

Lederer, A.L., Mirchandandi, D.A. & Sims, K. (1998). Using WISs to Enhance Competitiveness. *Communications of the ACM, 41* (7), 94-95.

Lee, S.M. & Chen, J.Q. (1997). A conceptual model for executive support systems. *Logistics Information Management, 10* (4), 154-159.

Leidner, D.E. & Elam, J. (1993). Executive information systems: Their impact on executive decision making. *Journal of Information Systems, 10* (3), 139-155.

Malhotra, Y. (2001). *Knowledge Management and Business Model Innovation*. Hershey, PA: Idea Group Publishing.

Mallach, E.G. (1994). *Understanding Decision Support Systems and Expert Systems*. Boston, MA: Irwin McGraw-Hill.

Mc Nurlin, B. & Sprague, R. (2002). *Information Systems Management in Practice,* 5th ed., New Jersey: Prentice Hall.

Copyright © 2003, Idea Group Inc. Copying or distributing in print or electronic forms without written permission of Idea Group Inc. is prohibited.

Messina, F.M. & Sanjay, S. (1995). Executive information systems: Not just for executives anymore! *Management Accounting, 77* (1), 60-63.

Millet, I. & Mawhinney, C.H. (1992). Executive information systems, a critical perspective. *Information & Management, 23* (2), 83-92.

Mintzberg, H. (1973). *The Nature of Managerial Work.* New York: Harper & Row.

Monash University. (1996). EIS Development Guidelines. *Technical Report 1/96,* Department of Information Systems, Faculty of Computing & Information Technology, Monash University, Victoria, Australia.

Mougayar, W. (1998). *Opening Digital Markets, Battle Plans and Business Strategies for Internet Commerce.* 2nd ed., New York: McGraw Hill.

Navarro, J.J. (1992). Computer supported self-managing teams. *Journal of Organizational Computing, 4* (3).

O'Brien, J.A. (1999). *Management Information Systems: Managing Information Technology in the Internetworked Enterprize.* 4th edition, Boston, MA: Irwin McGraw-Hill.

Palvia, P., Kumar, A., Kumar, N. & Hendon, R. (1996). Information requirements of a global EIS: An exploratory macro assessment. *Decision Support Systems, 16* (2), 169-179.

Poon, P. & Wagner, C. (2001). Critical success factors revisited: Success and failure cases of information systems for senior executives. *Decision Support Systems, 30* (4), 393-418.

Post, G. V. & Anderson, D.L. (1997). *Management Information Systems. Solving Business Problems with Information Technology.* Boston, MA: McGraw-Hill, 386-392.

PriceWaterhouseCoopers. (2000). *Technology Forecast: 2000.* California, USA.

Rai, A. & Bajwa, D. (1997). An empirical investigation into factors relating to the adoption of EIS: An analysis for collaboration and decision support. *Decision Sciences, 18* (1), 939-974.

Rainer, R.K., Snider, C.A. & Watson, H.J. (1992). The evolution of executive information system software. *Decision Support Systems, 8* (4), 333-241.

Rainer, R.K., Jr. & Watson, H.J. (1995a). The keys to executive information system success. *Journal of Management Information Systems, 12* (2), 83-98.

Rainer, R. K., Jr. & Watson, H. J. (1995b). What does it take for successful executive information systems? *Decision Support Systems, 14* (2), 147-156.

Roberts, B. (1995). Just team work. *PCWeek,* 4 Sept.

Copyright © 2003, Idea Group Inc. Copying or distributing in print or electronic forms without written permission of Idea Group Inc. is prohibited.

Rockart, J.F. & DeLong, D.W. (1988). *Executive Support Systems the Emergence of Top Management Computer Use*. Homewood, IL: Dow Jones-Irwin.

Ross, S.S. (2001). Get smart. *PC Magazine, 20* (14), 129-144, August.

Salmeron, J.L. (2001). EIS evolution in large Spanish businesses. *Information & Management, 1968*, 1-10.

Salmeron, J.L., Luna, P. & Martinez, F. J. (2001). Executive information systems in major companies: Spanish case study. *Computer Standards & Interfaces, 23*, 195-207.

Senn, J.A. (1996). Capitalization on electronic commerce. *Information Systems Management*, Summer.

Sprague, R. H., Jr. & Watson, H. J. (1996). *Decision Support for Management*. Upper Saddle River, New Jersey: Prentice-Hall.

Srivihok, A. (1998). Effective Management of Executive Information Systems Implementations: A Framework and a Model of Successful EIS Implementation. Ph.D. dissertation. Central University, Rockhampton, Australia.

Tang, H., Lee, S. & Yen, D. (1997). An investigation on developing Web-based EIS. *Journal of CIS, 38* (2), 49-54.

Theoharakis, V. & Serpanos, D. M. (2002). *Enterprise Networking: Multilayer Switching and Applications*. Hershey, PA: Idea Group Publishing.

Turban, E. (2001). California State University, Long Beach and City University of Hong Kong, USA. *Personal Communication*, 7 October.

Turban, E. & Aronson, J. (1998). *Decision Support Systems and Intelligent Systems*. Upper Saddle River, New Jersey: Prentice-Hall, Inc.

Turban, E., McLean, E. & Wetherbe, J. (1999). *Information Technology for Management*. Chichester, New York: John Wiley & Sons, Inc.

Turban, E., Lee, J., King, D. & Chung, H.M. (2000). *Electronic Commerce: A Managerial Perspective*. Upper Saddle River, New Jersey: Prentice-Hall, Inc.

Volonino, L., Watson, H.J. & Robinson, S. (1995). Using EIS to respond to dynamic business conditions. *Decision Support Systems, 14* (2), 105-116.

Warmouth, M.T. & Yen, D. (1992). A detailed analysis of Executive Information Systems. *International Journal of Information Management, 12* (2), 192-208.

Watson, H. J., Boyd-Wilson, T. & Magal, S. R. (1987). The Evaluation of DSS Groups. *Information & Management, 12* (2), 79-86.

Watson, H.J., Rainer, R.K. & Koh, C.E. (1991). Executive Information Systems: A Framework for development and a Survey of Current Practices. *MIS Quarterly, 15* (1), 13-30.

Copyright © 2003, Idea Group Inc. Copying or distributing in print or electronic forms without written permission of Idea Group Inc. is prohibited.

Watson, H.J., Houdeshel, G. & Rainer, R.K. Jr. (1997). *Building Executive Information Systems and other Decision Support Applications*. New York: John Wiley & Sons, Inc.

Westland, J.C. & Walls, J.G. (1991). Theoretical foundations for the design of Executive Systems in equivocal environments. *Proceedings of the Twenty-Fourth Annual Hawaii International Conference on System Sciences (HICSS-24)*, Hawaii, USA, 135-144.

Wheeler, F.P., Chang, S.H. & Thomas, R.J. (1993). Moving from an executive information system to everyone's information system: Lessons from a case study. *Journal of Information Technology, 8*, 177-183.

Whymark, G. (1991). Success in the implementation of an executive information system. *Proceedings of the Academy of International Business Southeast Asia Conference*, Singapore, 156_162.

Wreden, N. (1997). Business Boosting Technologies. *Beyond Computing*, November/December.

Worthington-Smith, R. (ed.) (2000). *The e-Commerce Handbook 2000. Your Guide to the Internet revolution and the future of business*. Trialogue, Cape Town, South Africa.

Worthington-Smith, R. (ed.) (2001). *The e-Business Handbook 2001. The 2001 Review of How South African Companies are Making the Internet work*. Trialogue, Cape Town, South Africa.

Copyright © 2003, Idea Group Inc. Copying or distributing in print or electronic forms without written permission of Idea Group Inc. is prohibited.

Chapter VIII

# SMEs in South Africa:
# Acceptance and Adoption of
# E-Commerce

Eric Cloete
University of Cape Town, South Africa

## ABSTRACT

*The chapter reports on research that was done in information systems at the University of Cape Town on the benefits of Internet utilisation and the barriers to its adoption amongst small businesses in South Africa. It addresses how these small businesses in a developing country perceive the potential benefits of e-commerce and look at their consequent adoption of e-commerce activities in their own organizations. Comparisons are made between studies conducted in first world countries, particularly regarding the role of government initiatives. A secondary aim of this research was to determine the current level of e-commerce adoption by small businesses in South Africa. This was achieved by circulating a questionnaire to test the perception of e-commerce benefits and e-commerce adoption levels amongst small businesses from various sectors in South Africa. If the global usage of the Internet for electronic commerce by small businesses is compared to the South African situation, this research clearly indicates that the available technologies are not adopted to the extent that is necessary for survival in a rapidly changing environment.*

Copyright © 2003, Idea Group Inc. Copying or distributing in print or electronic forms without written permission of Idea Group Inc. is prohibited.

# INTRODUCTION

This chapter reports on research that was done in Information Systems at the University of Cape Town on the benefits of Internet utilization and the barriers to its adoption amongst small businesses in South Africa. It addresses how these small businesses in a developing country perceive the potential benefits of e-commerce and look at their consequent adoption of e-commerce activities in their own organizations. Comparisons are made between studies conducted in first world countries, particularly regarding the role of government initiatives.

A secondary aim of this research was to determine the current level of e-commerce adoption by small businesses in South Africa. This was achieved by circulating a questionnaire to test the perception of e-commerce benefits and e-commerce adoption levels amongst small businesses from various sectors in South Africa.

If the global usage of the Internet for electronic commerce by small businesses is compared to the South African situation, this research clearly indicates that the available technologies are not adopted to the extent that is necessary for survival in a rapidly changing environment.

One of the most important features of using the World Wide Web is definitely the ability to conduct business over the Internet. The evolution of the Internet from a military tool to one that is available to organizations and individuals has opened many ways of doing e-commerce for SMEs (Small to Medium Enterprises). With the benefit of using the Internet to cut costs, improving efficiency and reach a much wider market, productivity and profit margins can be improved dramatically. Unfortunately, many South African SMEs are not utilizing the Internet adequately and therefore numerous excellent and unique business opportunities are never discovered or exploited for commercial gain (Courtney & Finch, 2001).

Although the classification of enterprises is normally based on size, turnover and number of employees, the definition of an SME differs in the literature; therefore, no fixed definition exists. An SME for the purposes of this research will be one that complies with the requirements of the South African National Small Business Act, No.102 of 1996.

E-commerce can be defined as the buying and selling of information, products and services with the assistance of computer technology and the Internet (Greenstein et al., 2000). This basically involves the exchange of electronic information between parties, normally followed by the exchange of goods and payment transactions. In the conduct of commerce, many differing activities might occur, such as marketing, interaction with clients and suppliers, interaction with government and acquisition products and the sales forthcoming of these events. Akkeren and Cavaye (1999) state that e-commerce improves an SMEs ability to compete with larger organiza-

Copyright © 2003, Idea Group Inc. Copying or distributing in print or electronic forms without written permission of Idea Group Inc. is prohibited.

tions and operate on an international scale. They also see e-commerce as a tool for providing cost effective ways for SMEs to market their business, launch new products, improve communications, gather information and identify potential business partners.

South Africa, as a developing country, faces many problems such as high unemployment, low levels of working skills, poverty and rampant crime. The encouragement of the development of SMEs might help to elevate these problems and improve the levels of skills in the country. This should contribute towards economic growth and assist in reducing poverty and unemployment levels in general. Information and communication skills have been proposed as means of increasing the productivity levels of SME's (Harrison et al., 1997). Yet the perception, implementation and the utilization of the Internet by South African SMEs is still relatively unknown (Cloete, 1999).

Electronic business, as an area of research, is still evolving with many researchers disagreeing to the benefits or boundaries. Many benefits from the utilization of e-commerce for SMEs are suggested (Davies, 2002), however its adoption seems to be low for a variety of reasons that are reported in this paper. Research performed on SMEs since the dotcom crash in 2000, indicates that perceptions and attitudes towards e-commerce have changed to quite an extent and people often look at potential benefits in a more pessimistic manner (Courtney & Finch, 2001).

## The Internet and SMEs

Many SMEs are forced into using electronic business due to their dominant clients forcing them to adopt or depart when the client implements an electronic purchasing system. The result is that SMEs that do not keep up with the e-evolution are in danger of being left out of tenders, particularly where larger corporations have electronic tender processes that might be attractive to SMEs.

Poon and Swatman (1997) identified short-term and long-term benefits for companies using e-commerce. Short-term benefits should be realized within months whereas long-term benefits may take longer and are normally fairly unpredictable. Their research showed that SMEs are not reaping significant short-term benefits from e-commerce and for those that did, the benefits were marginal and inconsistent. They identified long-term benefits as being the key motive for ongoing Internet activities.

Chan and Swatman (2001) found that the Internet as an e-commerce medium for EDI transactions is cheaper than traditional EDI, even after all the costs of implementing an Internet EDI solution was taken into account. They suggest the following Internet based B2B application as alternatives to traditional EDI:

Copyright © 2003, Idea Group Inc. Copying or distributing in print or electronic forms without written permission of Idea Group Inc. is prohibited.

- **EDI formatted documents over the Internet** (the use of FTP or e-mail applications to place the EDI messages into formatted documents and transmit them to the intended recipients)
- **XML EDI or XML document exchange** (the latest development in EDI translation, which aims to lower the barriers to e-business for SMEs).
- **Web-based EDI** (use of a web based form or intelligent interactive form to exchange business over the Internet).

Due to the ownership and decision making power of SMEs being held by one or two people, adoption of e-commerce into their organization is heavily reliant on these people's acceptance of the technology. The following factors are primary to the acceptance of e-commerce (Perry et al., 2001):

- **Usefulness** - The individual must be convinced of the relevant advantage of using e-commerce. Embracing e-commerce should be perceived as easier, faster, and cheaper than the current manual processes of doing business. E-commerce would need to be understood and considered to be useful by the individual.
- **Intention** - The individual must intend embracing e-commerce. They must have a positive attitude towards it. A person that views e-commerce as a tool that has no potential role in what they are doing is unlikely to accept the potential of e-commerce.
- **Ease of Use** - The individual must be comfortable with the use of technology. A person that is frightened to embrace technology will not be willing to accept e-commerce as a trading tool. If the appropriate skills and understanding of the technology are in place, then the use of e-commerce will be easier, thus making its acceptance more likely.
- **External Variables** - These are the factors affecting a person's acceptance of the technology that is beyond the control of the individual.

It is important to recognize that e-commerce activities range from entry-level activities such as having web browsers, web sites, and email, to sophisticated activities such as online payments, making purchases online, customer services and video conferencing. Akkeren and Cavaye (1999) state that the adoption of e-commerce practices is a progression and sophisticated technologies are unlikely to be adopted before those at the entry level have been successfully adopted. These entry-level activities provide the necessary technological infrastructure from which more sophisticated e-commerce activities can be developed.

According to the predictions of the Gartner Group, average European SMEs with ICT budgets of the order of $150,000 per annum would be unable to afford

Copyright © 2003, Idea Group Inc. Copying or distributing in print or electronic forms without written permission of Idea Group Inc. is prohibited.

an e-business application. According to Mandel (1999), SMEs in the USA are more likely to utilize Internet business strategies than their counterparts in Europe, due to the large homogeneous U.S. market. A 'static', advertising Internet website, for an SME in South Africa can start at the order of £10, 000 (Trialogue, 2000). This cost covers essentially marketing presence, with prices increasing as functionality is built in.

As an organization progresses up the ladder it must undergo change and become more sophisticated in its use of technology. While business implements these new changes, it is ultimately able to improve business efficiencies. The five progressive e-steps that a business may potentially advance through are summarized by Courtney and Finch (2001) as follows:

1. **E-mail:** This is defined as the use of e-mail to send messages, either to provide internal communication between staff, or to allow communication between businesses and their suppliers and customers. In e-commerce, the focus is on the external use of e-mail. Some elements of Electronic Data Interchange (EDI) can also be regarded as messaging, such as sending a request for a quotation.

2. **Website:** This is the establishment by a business of a website or e-mail list to publish information about products and services, so that their customers can access this information online. It gives the business an opportunity to create a greater awareness of its products to its customers and it places the business into a worldwide market. The website can publish marketing information, prices, and stock levels. In relation to supply-side activities, this is the use of other businesses' websites to identify suppliers and to acquire information on products and services. Checking the availability of products and services can also be done online; 24 hours a day, seven days a week.

3. **E-commerce:** This is the online interaction between a business and its customers, or a business and its suppliers, for the placement of an order. Online activities include issuing or receiving an invoice and an electronic payment.

4. **E-business:** This is the use of e-commerce to support the business relationship between a customer and a supplier, for example through the provision of interactive order progress tracking or online support. An integration of the supply chain links suppliers, manufacturing, and delivery, thus improving efficiencies and minimizing waste.

5. **Transformed Organizations:** The final outcome is the integration of all these activities with the internal processes of a business. The focus is customer service.

Copyright © 2003, Idea Group Inc. Copying or distributing in print or electronic forms without written permission of Idea Group Inc. is prohibited.

## Factors that Affect E-Commerce Adoption

The factors that affect e-commerce adoption are useful in determining the reason why an SME is at a certain level. These adoption factors are a result of the owner's business outlook as well as the organization's characteristics. Courtney and Fintz (2001) summarizes as follows:

- **Owner's Characteristics**

    Adoption of e-commerce is heavily reliant on the acceptance of e-commerce technology by the business owner. If the owner does not perceive the technology to be useful, nor understand its potential, then he/she will be reluctant to adopt it. The level of computer literacy of the owner and a lack of knowledge on how to use the technology will result in the business being less likely to adopt e-commerce (Kirby & Turner, 1993). If the owner is subjective and refers to the opinions of experienced people who recommend the adoption of e-commerce into the organization, then he is also more likely to accept their opinions (Harrison et al., 1997). SME owners are also concerned with return on investment. The pressure to show a return often leads to small firms being more concerned with medium-term survival rather than long-term viability (Akkeren & Cavaye, 1999). As a result, owners are often hesitant to make substantial investments when short-term returns are not guaranteed.

- **Organization's Characteristics**

    The amount of technology currently in use in the organization, such as PCs with modems and the use of email can make adoption easier (Iacovau et al., 1995). Businesses might adopt the use of e-commerce as a result of their competitors using it, so as not to lose their competitive advantage. If an organization has large amounts of data and transactions it is likely to influence its decision to adopt IT as this can help streamline operations and offer process efficiencies within the organization (Thong & Yap, 1995).

    In their research on small businesses, Akkeren and Cavaye (1999) found two factors affecting IT adoption that had previously not been recognized. The first was mistrust of the IT industry as some owners perceived the IT industry to be 'over-selling' the benefits of technologies and misinforming them. The second factor was a lack of time to get acquainted with the opportunities and challenges of the Internet due to its quick evolution rate.

Copyright © 2003, Idea Group Inc. Copying or distributing in print or electronic forms without written permission of Idea Group Inc. is prohibited.

- **Contextual characteristics**

The economic benefits of moving business transactions from fax, telephone and post to the Internet are well documented in many publications (Davies, 2002). Wilde and Swatman (2000) also noted that the forces of economic rationalism and globalism have enhanced the market as the final arbiter of price and service with the balance of power tilting from the manufacturer towards the consumer. Given this erosion of margins, companies need to reduce costs, both in production and transaction, in order to make their products and services more competitive. This again points to the Internet as a vehicle to reduce costs and to assist in obtaining a competitive advantage in the short term.

The traditional value chain has become virtualized to a great extent due to the fact that users of the Internet are able to order products and services online, without intervention of the purchasing department, while payment is made electronically using electronic funds or purchase cards. The primary activities in the Porter Value Chain, namely incoming logistics, outgoing logistics, marketing and sales, are being redefined in terms of how they are carried out and interact with each other, as technology provides for more sophisticated methods of business interactions (Walton & Miller, 1995; Porter, 1985).

These chains have become virtualized as the Internet was increasingly used as a '*binding agent*' (Davies, 2002). John Dobbs of Cambridge Technology partners describes '*value chain integration*' as a process of collaboration that optimizes all internal and external activities involved in delivering greater perceived value to the ultimate consumer (Economist, 1999). In the process, whole portions of the previous chain are being removed, redefined or disintermediated.

It is, however, necessary to note that the integration of value chains does not solve all problems. JIT (Just in Time) production is a methodology of reducing inventory stock that has been used for many years and serves as an example to prove this point. Critics of this method have pointed out that it merely forces the lower level manufacturer to hold stock and deliver to the client as needed. This method does however reduce stock holding costs throughout the higher levels of the value chain. It is also interesting to note that the Internet plays an increasingly important role with JIT ordering due to the ease of ordering at short notice.

Copyright © 2003, Idea Group Inc. Copying or distributing in print or electronic forms without written permission of Idea Group Inc. is prohibited.

Researchers have further identified a number of additional issues that impact the adoption of e-commerce. These are summarized by Courtney and Fintz, 2001:

- *Low use of e-commerce by Customers and Suppliers.* This means that there is little incentive for SME's to engage e-commerce until their customers and suppliers are also using it.
- *Concerns about security aspects.*
- Concerns about legal and liability aspects.
- *High costs of development and computer and networking technologies commerce.*
- *Limited knowledge of e-commerce models and methodologies.*
- *Unconvinced of benefits to the company.*

The South African SME business sector in 1996 was estimated to number in the order of 800,000 (Darrol, 1996). Viviers and Sootinis (1999) suggest that this group accounts for approximately 46 percent of the total South African economic activity.

## Survey Results

The study by Davies (2002) concentrated on trade, manufacturing and construction sectors, mainly in the Johannesburg-Pretoria area, while the study by Courtney and Fintz (2001) was aimed at the manufacturing sector in the Western Cape.

Both surveys encountered low response rates, mostly due to resistance and to a certain extent unhelpfulness by potential survey companies. While both surveys supported the findings of the other, only the study by Davies will be commented on in this paper.

- **ICT usage**

Two hundred and fifty three SMEs were targeted from the Gaffney's Business Directory, Gaffney (2001), with a response rate of 19.4 percent. The first part of the survey, questions one to four, determined the economic sector and ICT utilization. In the response, 80 percent of the organizations that returned survey forms (39) claimed their PCs were connected to the Internet and 35 servers were utilized, with ten connected to the Internet. Only five mainframes were used, of which one was connected to the Internet. One company used a total of six stand-alone Point of Sale devices, with no Internet connection. Six notebooks were listed by the 49 organizations, with only two of these connected to the Internet. Two companies claimed to have no Internet access and one possessed no computer facilities. A total of 20 organizations had websites, with a mere five utilizing EDI.

Copyright © 2003, Idea Group Inc. Copying or distributing in print or electronic forms without written permission of Idea Group Inc. is prohibited.

*Table 1: Perception of Industry Participant's Utilization of the Internet*

| Competitors | Clients | Partners | Suppliers |
|---|---|---|---|
| 71% | 78% | 31% | 61% |

- **Level of Internet usage**

  Questions five to seven attempted to discover the level of Internet usage. The 49 organizations' perception of Internet usage is shown in Table 1.

  Question seven determined the type of Internet usage by the respondents:

*Table 2: Responses to Question 7 of Questionnaire*

| Percentage of respondents | Q7: What does the organization utilize the Internet for? |
|---|---|
| 71% | Electronic mail |
| 69% | Document transferring |
| 69% | Financial: paying bills, salaries, invoicing, etc. |
| 57% | B2B transactions |
| 49% | Marketing |
| 41% | Submitting tenders to customers |
| 37% | Purchasing raw materials, office supplies, etc. |
| 35% | Making order information available to customers |
| 29% | Order placement by customers |
| 27% | B2C transactions |
| 16% | Interaction with government |
| 16% | Link into outside computerized systems to perform orders |
| 10% | B2Gov transactions |
| 4% | Voice/Audio communication (VOIP) |
| 0% | Video conferencing |

- **Perceptions of the Internet**

  Questions 8 to 20 were focused on the effect of the Internet and the perceptions of the Internet. The values 1-4 were divided into four options: strongly disagree (1-1.6), disagree (1.61-2.2), neutral (2.21-2.8), agree (2.81-3.4) and strongly agree (3.41-4). A similar grading method was used for values on the 1-5 scale.

- **Reasons for not using the Internet**

  Question 21 was aimed to identify the top ten barriers for non-Internet usage.

Copyright © 2003, Idea Group Inc. Copying or distributing in print or electronic forms without written permission of Idea Group Inc. is prohibited.

*Table 3: Responses to Questions 8-15, 17-20*

| Question No. | Q 21: Barriers for non-Internet usage | Response |
|---|---|---|
| 8 | The I can assist: making my organization more efficient/reduce costs | Agree |
| 9 | The I is adequately used by my organization for our bus functions | Neutral |
| 10 | The I is able to assist in the type of business my organization operates with | Agree |
| 11 | More usage of I could help make my organization more competitive. | Agree |
| 12 | Our knowledge of the I is sufficient for our business | Neutral |
| 13 | Our supplier/customers do not require I interaction from us | Disagree |
| 14 | My customers are using the I to order products/services in our industry | Neutral |
| 15 | My competitors are using the I to improve their competitiveness | Agree |
| 17 | The I enables us to manage our business more effectively | Neutral |
| 18 | The I has altered the way in which we do business | Agree |
| 19 | The I enables us to improve our customer service | Neutral |
| 20 | The I enables us to improve our supplier relationships | Neutral |

*Table 4: Top 10 Barriers Discouraging Internet Usage*

| Ranking | Q 21: Barriers for non-Internet usage |
|---|---|
| 1 | Lack of time to investigate options |
| 2 | Lack of knowledgeable/qualified staff |
| 3 | Lack of information on options |
| 4 | Not enough business partners use it |
| 5 | No real business need/no benefit |
| 6 | Lack of HR & skills |
| 7 | Poor telecommunications infrastructure |
| 8 | Lack of training on e-commerce |
| 9 | Poor security on the Internet/fraud |
| 10 | Costs too high |

Copyright © 2003, Idea Group Inc. Copying or distributing in print or electronic forms without written permission of Idea Group Inc. is prohibited.

- **Problems experienced by those using the Internet**

Question 22 attempted to identify the problems experienced by those who claimed to be using the Internet for e-commerce:

*Table 5: Problems Experienced by Those Using the Internet*

| Ranking | Q 22: Problems experienced by those who used the Internet |
|---------|------------------------------------------------------------|
| 1 | Insufficient cheap bandwidth |
| 2 | Not enough companies use it |
| 3 | Technical barriers |
| 4 | Expensive hardware |
| 5 | Lack of e-commerce information |
| 6 | Immature e-commerce technologies |
| 7 | Lack of online guarantees |
| 8 | Lack of knowledgeable/qualified staff |
| 9 | Lack of import/export information |
| 10 | Legal issues |

- **Benefits experienced by those using the Internet**

Question 23 asked for the top ten benefits experienced as a result of Internet usage.

*Table 6: Benefits of Internet Usage*

| Ranking | Q 23: Benefits of Internet usage |
|---------|-----------------------------------|
| 1 | Financial reasons; such as online banking |
| 2 | Speed of communication/response rates |
| 3 | Reduction of communication costs |
| 4 | Research/gathering of information |
| 5 | Improve productivity/competitiveness |
| 6 | Global reach |
| 7 | Improve CRM |
| 8 | Transfer documents |
| 9 | Saves time |
| 10 | Marketing/advertising |

Copyright © 2003, Idea Group Inc. Copying or distributing in print or electronic forms without written permission of Idea Group Inc. is prohibited.

# CONCLUSION

The findings of this survey indicate that many issues from overseas research are both pertinent and valid for the South African environment. Information on the benefits of e-commerce utilization and the barriers to its adoption from this survey and the survey from Courtney and Fintz (2001) show some correlation to one another, as well as to the results obtained by overseas surveys. The Internet has been shown to be a disruptive technology that has influenced the business environment. Although respondents indicated that greater usage of the Internet would improve their businesses' competitiveness, costs of implementing and running e-commerce business are perceived to be strong barriers to greater adoption by South African SMEs.

In summary, it can be said that the following objectives were achieved by this research:

*   A good indication of the level of e-commerce adoption by SMEs in South Africa has been perceived.
*   The barriers to the adoption, implementation and utilization of e-commerce by South African SMEs have been highlighted.
*   The benefits of e-commerce adoption by South African SMEs have been pointed out.

The role of the government in the greater adoption of electronic commerce cannot be understated. The introduction of a competitor to Telkom is a very important factor if greater competition can reduce communication costs. If the new bill for e-commerce is to further introduce barriers to Internet usage, the South African government will need to reassess their policy towards e-commerce for South Africa to be competitive in the global market.

# REFERENCES

Akkeren, J. & Cavaye, A. (1999). *Factors Affecting the Adoption of E-commerce Technologies by Small Business in Australia – An Empirical Study*. Retrieved October 1999 from the World Wide Web: www.acs.org.au/act/events/io1999/akkern.html.

Chan, C. & Swatman, P. (2001). *Management and Business Issues for B2B e-Commerce Implementation*. Deakin University, Australia. Retrieved May 12, 2001 from the World Wide Web: http://mis.deakin.edu.au/research/working_papers_2001/2001_09_chanc.pdf.

Cloete, E. (2001) *E-commerce: A Contemporary View*. Pardus Publishing.

Copyright © 2003, Idea Group Inc. Copying or distributing in print or electronic forms without written permission of Idea Group Inc. is prohibited.

Cloete, H. (1999). *How Does Information Technology Influence Small Business Growth?* Technical Report to the Department of Information Systems, University of Cape Town.

Courtney, S. & Fintz J. (2001). Small Businesses' Acceptance and Adoption of E-commerce in the Western-Cape Province of South-Africa. Empirical Research Project, Department of Information Systems, UCT.

Darrol, S. (1996). A new era for small business? *Prodeer Newsletter, 8* (1), 4-17.

Davies, S. (2002). South African SME's and Internet Based Electronic Business. Empirical Research Project, Department of Information Systems, UCT.

Economist (1999). You'll never walk alone. *The Economist Newspaper,* Survey on business and the Internet, 26 June 1999.

Gaffney, P. (2001). *Gaffney's Business Contacts 2001.* The Chamber of Commerce and Industry Yearbook, The Gaffney Group, Johannesburg, South Africa, ISBN 0-9583821-2-3.

Greenstein, M. & Feinman, T.M. (2000). *Electronic Commerce.* Irwin/McGraw Hill.

Harrison, D.A. Mykytyn, P.P. & Rienenschneider, C.K. (1997). Executive decisions about it adoption in small business: theory and empirical tests. *Information Systems Research, A Journal of the Institute of Management Sciences, 8* (2) 171-195.

Iacovou, C.L., Benbasat, I., & Dexter A.A. (1995). Electronic data interchange and small organizations: Adoption and impact of technology. *MIS Quarterly, 19* (4), 465-485.

Kirby D. & Turner M. (1993). IT and the Small Retail Business. *International Journal of Retail and Distribution Management, 21* (7), 20-27.

Mandel M.J. (1999). The Internet economy: The world's next growth engine. *Business Week Magazine,* New York: McGraw-Hill, Oct., 44-49.

Perry, G. P. & Schneider, J.T. (2001). *Electronic Commerce.* Thomson Publishing.

Poon S. & Swatman P. (1999). An exploratory study of small business Internet commerce issues. *Information and Management, 35* (9-18).

Porter M. (1985). Competitive Advantage. Free Press.

Thong, J. & Yap, C. S. (1995) CEO characteristics, organizational characteristics, and information technology adoption in small business. *Omega,* August, *23* (4), 429-442.

Trialogue, (2000). *The E-commerce Handbook.* Trialogue Publications, Cape Town.

Copyright © 2003, Idea Group Inc. Copying or distributing in print or electronic forms without written permission of Idea Group Inc. is prohibited.

Viviers, W. & Sootinis, W. (1999). South African SMEs: Obstacles to Export to the SADC, Potchefstroom University. Retrieved June 20, 1999 from the World Wide Web: http://www.sbaeur.uca.edu/DOCS/98icsb/j011.htm.

Walton, L.W. & Miller, L.G. (1995). Moving toward is theory development: A framework of technology adoption within channels, *Journal of Business Logistics, 16* (2), 117-135.

Wilde, W.D. & Swatman, P.A. (2000). *Studying Telecommunications Enhanced Communities: An Economic Lens.* Deakin School of Management Systems, Victoria, Australia. Retrieved July 2001 from the World Wide Web: http://mis.deakin.edu.au/research/working_papers_2000/2000_07_wilde.pdf.

Copyright © 2003, Idea Group Inc. Copying or distributing in print or electronic forms without written permission of Idea Group Inc. is prohibited.

Chapter IX

# Key Indicators for Successful Internet Commerce: A South African Study

Sam Lubbe
Cape Technikon, South Africa

Shaun Pather
Cape Technikon, South Africa

## ABSTRACT

*The advent of e-commerce has brought about new implications on research directions in the business arena. Is it not sufficient to just study the formation of electronic markets in e-commerce. It is also necessary to have insight into the electronic markets' innermost workings. This chapter, therefore, highlights the factors that give these new organizational forms (e-commerce enterprises) sustainable competitive advantages. The chapter is present in three main sections: A study of variety of reports of practitioners and researchers from the international arena, provide a background of prior knowledge in the filed. The analysis of this knowledge base is conceptualized into a framework of key factors affecting e-commerce success. Finally, the results of an empirical study of South African e-businesses are reported on. The authors provide some insight on the application of the conceptual framework as applied in the South African situation.*

Copyright © 2003, Idea Group Inc. Copying or distributing in print or electronic forms without written permission of Idea Group Inc. is prohibited.

# INTRODUCTION

The world of Internet commerce has been rapidly evolving since its advent in the 1990s. This has had implications on research directions in the field of electronic commerce (e-commerce). No longer is it sufficient to study the formation of electronic markets in e-commerce. It is also necessary to have insight into the electronic markets' innermost workings. This chapter, therefore, highlights the factors that give these new organizational forms (e-commerce enterprises) sustainable competitive advantage, thus allowing them to create value in the Internet marketplace, and increase their ability to maximize profits.

The following sections review various factors relating to e-commerce successes and failures. The analysis of these factors provides, enabling directions in successful adaptations and new interpretations of long-standing issues that senior managers face with new directions of Internet commerce.

The chapter is presented in three main sections. First, a study of a variety of reports of practitioners and researchers from the international arena, provide a background of prior knowledge in the field. Then the analysis of this knowledge base is conceptualized into a framework of key factors affecting e-commerce success. Finally, the results of an empirical study of South African e-businesses are reported. The authors provide some insight on the application of the conceptual framework as applied in the South African situation.

*Figure 1: Types of E-Commerce*

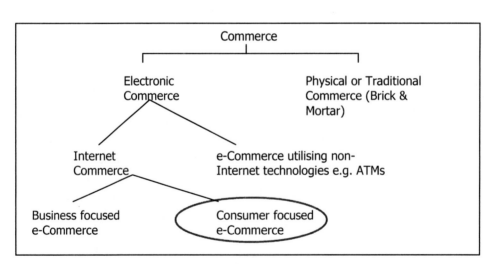

*Source: Adapted from Chan et al., 2001*

Copyright © 2003, Idea Group Inc. Copying or distributing in print or electronic forms without written permission of Idea Group Inc. is prohibited.

*Placing the Study in Perspective*

The term e-commerce conjures various interpretations. Figure 1 clarifies the context in which e-commerce is referred to in this chapter.

# A Review of Issues Affecting E-Commerce Adoption and Implementation

The following section reviews various reports of e-commerce in practice. The list of issues is not exhaustive, but rather an attempt to bring to the fore a varied number of issues that warrant attention.

## Adopting E-commerce: Business Benefits

A large number of reports comment on the benefits of moving a business to an e-commerce setup. Auger and Gallaugher (1997) noted the following benefits affecting the adoption of an internet-based sales presence for small businesses: First, it gives the business access to an affluent customer base and provides lower information dissemination costs. Secondly, transaction costs could be lower while a broader market can be reached. Thirdly, the e-commerce site should also help with additional channels for customer feedback, while at the same time it should enable consumer and market research.

Kalakota and Marcia (1999, cited in O'Brien: 2002) provide a good overview of why companies are building e-commerce web-sites: generation of new revenue from online sales; reduced costs through online sales and customer support; attracting new customers via Web marketing and advertising; increased loyalty of existing customers via improved Web customer service and support; developing new Web-based markets and distribution channels for existing products; and development of new information-based products accessible on the Web.

There may be many other reasons why companies are adopting e-commerce. However, whatever the reasons are, it is important to recognize factors that influence both success and failure.

## Problems Faced with E-commerce Adoption

In addition to the benefits highlighted above, many reports also provide insight to the problems that accompany e-commerce enterprises.

Wilson (2001) argued that many organizations fail because their venture capital firm got cold feet. He also noted that about one dot-com organization fails per day and many more are in trouble (failures like Pets.com and Garden.com are cited). Some mistakes that these organizations have made were: Too many competitors, short-term mentality, undisciplined growth, unrealistic revenue pro-

Copyright © 2003, Idea Group Inc. Copying or distributing in print or electronic forms without written permission of Idea Group Inc. is prohibited.

jections, inexperienced management, underestimating the costs of establishing a national brand and a lack of customer-centered focus.

Wilder (1998) noted that web commerce is changing the way a lot of organizations do business, but that it is not everything that has been hoped for. The e-commerce concept had been hyped and valuations of these company's shares had been high – all because of exaggeration and oversimplification of the real issues. Wilder (1998) provides a succinct list of e-commerce myths. The first misconception, highlighted by Wilder (1998), is that *everybody thinks that it is easy* (a.k.a., the barriers have never been lower). It is easy to put up a web site, but add words such as effective, scalable, successful, and it gets a lot harder. Web infrastructure could also be a risk for older organizations. The second myth is that it is cheap. Perhaps e-commerce is cheap when compared with a full-blown enterprise resource planning implementation. The third myth is that everybody is doing it. Wilder (1998) argues that there are successful companies on the sideline of e-commerce that include chains Best Buy and Fry's Electronics.

The fourth myth highlighted by Wilder (1998) is that it is lucrative. However, there are still several examples of businesses that do better at traditional retailing, e.g., The Burlington Coat Factory has a lower sale output through its Internet presence compared to the output from its 250 stores. The next myth is that the Web levels the playing field (a.k.a., startups can instantly compete on the same footing as long-established companies). Wilder argues that companies that want to be successful at web commerce need the marketing clout, brand identity, and scale to do back-end fulfillment and customer service – and the capital. This could be a possible reason why big-physical-world competitors had bought many start-ups.

Some interesting lessons came from an article in Business Inc. (2001) on the 100 dumbest moments in e-commerce history. They discussed the demise of firms such as Boo.Com, Beyond.com, etc. They also noted that Internet companies make thousands of mistakes every week. It was argued that adopters of e-commerce enterprises do not learn from these mistakes. Many of these organizations did not startup addressing the correct market or they incorrectly spent the money that was invested in them.

Auger and Gallaugher (1997) comment specifically on the problems faced by small businesses: difficulty monitoring the site use; security concerns; analyzing and promoting the site; and lack of access to expertise.

Being a player in the e-commerce domain means that the organization must become more responsive. It should have an agility in responding to environmental forces, which play themselves out much quicker than in brick and mortar setups. ITQuadrant (2001) refers to being able to execute e-commerce strategies rapidly. Organizations must be able to respond quickly to market changes, emerging

Copyright © 2003, Idea Group Inc. Copying or distributing in print or electronic forms without written permission of Idea Group Inc. is prohibited.

opportunities, and competitive threats. This implies that the e-commerce enterprise should be able to implement new business solutions quickly. This impacts on the implementation of applications, and, ITQuadrant (2001) asserts that this requires an adaptive application architecture that maximizes reuse.

Kandiah et al. (2000) in an exploratory study identified some key variables that contribute to IS project failure. Some of them are poorly communicated goals, lack of corporate leadership, inadequate skills, poor project management, deviation from timetable and budget, data collection strategies, sampling approach and principal component analysis. They argued that IS projects such as e-commerce often fall short of expectations.

*Systems Development*

Methodologies for developing information systems (IS) have matured considerably over time. Years of experience have culminated into various packages of good practice, and these are evidenced in a multitude of software engineering literature. There are, however, many common *do's and don'ts* advocated for sensible development of information systems. There is no reason that this should be thrown out of the window when developing information systems that use e-commerce as the primary driver. Bragg (in Highsmith et al., 2001) argues that worst practices in developing the e-commerce software itself vary widely, and many of them are no different from worst practices in developing other kinds of software. For example, carefully defining the business problem and scoping the system is advocated by many analysis and design authors (see Whitten et al., 2001; Satzinger et al., 2001). It is therefore not a new problem, because recent reports of e-commerce implementation problems have pointed to similar issues. Additionally, sticking to a sound methodology rooted in the systems development life cycle (SDLC) is one aspect that really should not be neglected. There are, however, alternate points of view. For example, Highsmith et al. (2001) claim that even though these new practices often fly in the face of conventional wisdom, they survive because they are better adapted to the new reality. They propose three essentials for e-commerce success: adaptive enterprises, adaptive software, and agile IT professionals.

When incorporating e-commerce platforms into existing business practices, various strategies could be used. For example, e-business reengineering is finally leading companies to consider the deep, radical redesigns that Hammer and other BPR gurus first advocated in the early 1990s (Harmon, 2002). What was radical then is now necessary for survival. He (Harmon, 2002) argued that a company cannot transition to an e-commerce strategy without creating a new company software architecture and an extensive new suite of software applications. To

Copyright © 2003, Idea Group Inc. Copying or distributing in print or electronic forms without written permission of Idea Group Inc. is prohibited.

succeed in e-commerce, these new software applications will be required to integrate the organization's existing databases, legacy applications, and business processes more extensively than anything ever attempted.

It is also important that proper information gathering techniques are employed to ensure that the requirements on the e-commerce system are carefully defined. Hayes (2002) noted that comprehensive requirements that drive bottom-up project plans resulting in reasonable, feasible schedules are rarely being achieved. Most of the schedules she has scanned are unrealistic. It is suggested that the reason for this shambles is that software is being used as a competitive weapon and is driven by the need for e-commerce. She noted that back office to frontline meant something for the managers. Organizations realized that software systems could differentiate them in the marketplace. This meant new functionality because a new product could be created within days. There is no manual alternative, and delays might mean missing market share.

## Planning the E-Commerce Venture

Good advice is provided by Abels (2002), who argues that the first step toward success in e-commerce is to throw aside the baggage of the industrial age and adopt a born on the Web business model and an Internet mindset.

Pickering (2001) noted in an article on a survey of the e-commerce and IT practices that the right reasons should be picked when starting out. They caution, however, that these reasons might not be the right ones at the beginning. Wilson (2001) provides other advice pertinent to the planning phase, suggesting that businesses should select online target markets with clearly unfilled or under-filled niches. While it is advocated that the short-term goals had to be clearly defined, plans should also focus on the mid- to long term.

Financial feasibility is another traditional planning practice that also deserves attention. Wilson (2001) advises dot-com startups to set realistic revenue and margin projections. He suggests that many dot-com "dreamers" set high and unrealistic projections – they should instead re-evaluate the information and services they are giving away for free.

Wilson (2001) provides insight into two further lessons learned from analyzing e-commerce failures. Those businesses that lack sufficient experience and expertise in an e-commerce setup are encouraged to obtain expert advice as inexperience sometimes results in serious planning and execution errors, e.g., trying to patch traditional structures onto a new type of company. Secondly, planning should not obviate the customer's important role. An unremitting customer focus is demanded. Many of the dot-coms' ideas are unrelated to real customer demand.

Copyright © 2003, Idea Group Inc. Copying or distributing in print or electronic forms without written permission of Idea Group Inc. is prohibited.

*Testing and Training*

Furthermore, because e-commerce incorporates more than one discipline, a variety of consultants are used - and they do not solve everything (James, 1997). Most of the time, there are no contingency plans and systems are not tested. Many people do not train their staff and organizations, then enter the denial stage.

## Involvement of Stakeholders (Top Management)

Involvement of a range of stakeholders, from the end-user right through to top management is one of the key issues in systems development. In fact, the Standish group report stated (see Johnson, 1995) that lack of users' involvement was a key problem contributing to IS failures. As already suggested, e-commerce development, should be rooted in sound methodology. James (1997) provides an example of a departure from this. He suggests that one of the big problems with the development of e-commerce sites is that top management is clueless, especially when business goals are fuzzy.

One non-negotiable role of involvement of the top management is alignment of e-commerce projects with organizational objectives (Abels, 2002). This strategic function bears an important relevance on steering transition in the right direction. Furthermore, it should be the role of top managers to allocate responsibility for the development of the site. The responsibility for the development of the site, according to Auger and Gallaugher (1997), should come from inside the organization. However, Abels (2002) suggests that the e-commerce project team should be created separately from the organization's IT department, and should include marketing strategists form the beginning.

ITQuadrant (2001) reaffirms the importance of senior management's role. They noted that senior management must be heavily involved in the development of the company's e-commerce direction right from the start. The e-commerce strategy must be tied to the business plan. E-commerce initiatives must also be well integrated with the overall business strategy.

Consideration to the needs of end-users should also not be neglected. Traditional thinking is too rigid and internally focused and fails to take into account the goals of the user, according to Abels (2002). James (1997) in reporting on various e-commerce "fiascoes" that prevail also refers to end-users' involvement. One of several issues commented on is that the people that use the site (end-users) do not participate in the development. Projects can fail if they do not meet user needs. Once again, this points to e-business's neglect of well-researched and tested systems development methodology.

Abel (2002) asserts that users want fast, friendly, easy-to-use solutions. Developers should learn users' goals. The site should be structured for Web users

Copyright © 2003, Idea Group Inc. Copying or distributing in print or electronic forms without written permission of Idea Group Inc. is prohibited.

or the organization will not win. He also mentions that failing to test early, often, realistically and outside the organizational intranet is an expensive mistake. Lastly, Abel (2002) suggests that attention must be paid to format of the content. Organizations should remember that readers normally scan for certain key words. People most of the times visit a site for reasons not anticipated.

## Marketing and Choice of Products

The marketing function is one aspect that has been misinterpreted by e-commerce businesses. Highsmith et al. (2001) suggest that the most profound of recent changes are not in technology but in the relationship between the business and its external markets and customers.

Wilder (1998) reports that one of the myths perpetuated by the advent of e-commerce is that it means the end of mass marketing. The Web is the first communications channel that enables cost-effective one-to-one marketing on a huge scale. Brand names are a key issue. Mass marketing is therefore a necessity for the leaders of the online industry.

Another lesson is to plan an aggressive marketing budget with realistic goals. One of the chief errors committed by failed dot-coms was based on the belief that they could buy a national brand. This only takes place over time and demands excellence in customer service, product fulfillment, and product quality (Wilson, 2001).

When strategizing e-commerce concepts, a business seeking new grounds should consider the appropriateness of the product for the Web (Auger & Gallaugher, 1997). Wilder (1998) in writing about e-commerce myths, claims that business believes that it leads to product commodization. Price is not the number one selling point for most companies online. Customers also want brands and service they can trust.

## Adopting a New Culture of Doing Business

The move from a traditional brick and mortar setup to an e-commerce enterprise can be a test of the organizations ability to adapt to a new way of business. One of the important factors in such a transition is the development and cultivation of an e-commerce culture (ITQuadrant, 2001). Many organizations have reported that cultural change is one of the top challenges in moving to an e-commerce enterprise. Perhaps an assessment of the cultural feasibility of the project would assist in this regard (see Satzinger et al., 2001). In this transitional period, the e-commerce organization should note the various stages of human and organizational assimilation of new technologies (Erwin & Blewitt, 1998).

Copyright © 2003, Idea Group Inc. Copying or distributing in print or electronic forms without written permission of Idea Group Inc. is prohibited.

## Keeping the E-Business Operational

InterOPS Management Solutions (2001) asserts that the key factor for e-commerce operations management is that e-commerce operations require its own protection plan. Relocated operations must also continue to be monitored and managed to ensure that they are fulfilling all functional requirements. The last key factor they mentioned is that e-commerce continuity procedures must be testable to verify their readiness.

According to them, there are some benefits to a smart e-commerce operations management strategy. These benefits are highly cost-efficient risk mitigation, additional value during normal operations and third shift management (ensuring 24 hour/7 day coverage). The last benefit is that there could be customer and partner risk reduction.

## Supply Chain Management

Oakes (2002) noted that the current vision of Chief Information Officers is to forge ahead without the full picture of what supply chain management (SCM) can do. He identified some myths; the first one was implementing SCM solutions always creates benefits. He notes that only realistic and diligent Return on Investment (ROI) can ensure this. The second myth is that the Internet has dramatically reduced supply-chain cycle times. These times could still be further reduced when third-party providers start using SCM technology.

Another myth according to Oakes (2002) is that SCM portals are just another short-term fad. The problem is that the word *portal* has been overused to the point of it becoming meaningless. The argument is that smaller communities could belong to multiple supply-chain communities. The fourth myth is that everything inside the four walls works just fine. There is no synchronization across departments or with trading partners. The fifth myth cited is that investing in SCM applications will always result in a rapid return. The promises of application vendors can only be realized through focus on delivery, quality and time targets. Promises should not be brought into the equation above, but should be realized through focus on delivery, quality and time targets.

## Steps to Launching the Business

Wilson (1999) noted that there are several steps necessary to start a successful new online business. The first step is to develop the concept as clearly as possible. Another step is to prepare a unique selling proposition – refine your concept down. A clear unique selling position is an essential component for a prosperous e-commerce venture. The next step is to obtain a secure domain name. Ensure that the name is not too difficult to spell and that people would be able to remember it.

Copyright © 2003, Idea Group Inc. Copying or distributing in print or electronic forms without written permission of Idea Group Inc. is prohibited.

The next step is to secure funding for the project; sometimes this can forestall going any further. A gradual approach to start-up is called for most of the time. The next step would be to construct the website and e-mail systems. The last step would be to execute marketing strategies.

## Research Methodology

The foregoing sections have highlighted various factors that impact implementation and adoption of e-commerce. Several issues, both positive and negative, were examined. In this section, the research methodology that will be applied in analyzing the literature is discussed.

### Aims of the Research

The aims of the research are:
- to identify, study and evaluate e-commerce success factors;
- to develop a preliminary theory of good practice in the field of e-commerce;
- to test this theory or theoretical conjecture by reference to South African organizations and practitioners; and thereby,
- develop the theory into managerial guidelines.

### Research Options

The research has a theoretical underpinning from the literature review. This was the result of the study of the research area through the writings of others and through discourse with learned or informed individuals. The theory produced a list of success factors through content analysis (See Fig. 2) and these would be tested in practice. These factors would also be tested using a questionnaire based on ranking and on preference using a seven-point Likert scale.

### Analysis of Theoretical Evidence

The main approach to the analysis of evidence obtained from the literature was to use content analysis in the pilot study as described by Huberman & Miles (1994). Because content analysis was used, a positivist approach was followed by a realist and interpretive approach to discuss the results and interprets in order to apply the conclusions. These techniques require the researcher to carefully comb through the evidence collected from the literature in order to find patterns that display general principles applying to e-commerce. Because of detailed work, empirical generalizations and a theory could be developed.

Copyright © 2003, Idea Group Inc. Copying or distributing in print or electronic forms without written permission of Idea Group Inc. is prohibited.

# Data Collection

## South African Background

A brief background to the South African situation is presented, as the theoretical conjecture described through content analysis will be tested during phase two of this study, with e-commerce organizations based in this country.

South Africa is a medium-sized country, an area of 471,000 square miles at the southern tip of the African continent, with a population of some 44 million people. It is relatively industrialized in comparison with the rest of Africa. South Africa is a wealthy country from an industrial and agricultural point of view and computers have been actively in use in South African business and industry since the early 1960s when both IBM and ICL opened offices in Johannesburg. Today, South Africa employs computers in every aspect of industry, business and government as well as having a relatively high percentage of home computers among the middle and upper class. All the major vendors are present and there is considerable interest in the use of Internet technologies to support business.

## The Development of the Theory

After summarizing the theory, it became necessary to analyze the data that had been collected in order to see what general principles relate to the assessment of success factors. The main process that was used was content analysis. Content analysis may be defined as the process of determining the fuller, detailed meaning of a document, manuscript, speech or any type of communication in a way that is both reliable and replicable. A formal definition of content analysis is supplied by Berelson (1952) as a research technique for the objective, systematic and quantitative description of the manifest content of communications. To perform content analysis, one must investigate the frequency and intensity of which concepts are addressed in the text. It is a simple but laborious process, since one must closely examine the documentation for concepts, particularly those that are repeated often. Berelson (1952) noted that content analysis might be used with different units of analysis including words, themes, characters, items and space-time measurements.

The results of the content analysis were summarized into a main set of factors that could be regarded as success factors for e-commerce. Frequency tables showing the number of occasions in the discussion of the main themes that occurred during the analysis of the articles were drawn up. In this study, only themes that were raised by informants more than three times were included in the frequency table.

In Figure 3, a list of the themes raised by the theory and their frequency are shown. The frequency table is based on the analysis as described in the previous section. The nature of the drivers is different as some of the drivers originate from the upper echelons of management (e.g., alignment of the goals) while some

Copyright © 2003, Idea Group Inc. Copying or distributing in print or electronic forms without written permission of Idea Group Inc. is prohibited.

originate from the bottom of the business chain (e.g., feedback). It should also be mentioned that the authors specifically named each group. It must also be stressed that the literature review is by no means exhaustive and can easily be extended.

Figure 3 shows that in the theory there were 17 major references to the importance of conducting research and development during the conversion to an e-commerce business and 15 references to financial planning of e-commerce as well as 15 references to feedback from staff and customers as a condition to ensure e-commerce success, etc. These success factors are described in the following section.

## Discussion of Themes Identified

*Research, Development and Skilling of the e-commerce site:* This theme could be described as the impact of doing proper research on what is required of an e-commerce site. Proper research can ensure that the organization can keep its market share by employing the key attributes of e-commerce. An e-commerce team that is not part of the IT department should do the development of the site. The team should maximize the flow of ideas and educate people about new processes, systems and business practice.

*Capitalization:* The informants viewed this as important. Organizations should keep as close as possible to the budget and ensure that they do not overspend or run short of money. E-commerce enterprises should set realistic revenue and margin projections. The budget should ensure that there is enough

*Figure 2: Content Analysis Showing Relative Frequencies and Percentages for E-Commerce Success Factors*

|  | Themes |  | Total | % | Accum % |
|---|---|---|---|---|---|
| 1 | Research, Development and skilling of the site | RD | 17 | 15 | 15 |
| 2 | Capitalisation | FA | 15 | 12 | 27 |
| 3 | Feedback | FB | 15 | 12 | 39 |
| 4 | Management | MM | 14 | 11 | 50 |
| 5 | Marketing | MG | 13 | 10 | 60 |
| 6 | People (staff and customer) | SC | 12 | 10 | 70 |
| 7 | Alignment of Goals | AG | 12 | 10 | 80 |
| 8 | Items not suitable for the Internet | IN | 11 | 9 | 89 |
| 9 | Parameters | PM | 8 | 8 | 97 |
| 10 | Security | SC | 4 | 3 | 100 |
|  | **Total** |  | **121** |  |  |

Copyright © 2003, Idea Group Inc. Copying or distributing in print or electronic forms without written permission of Idea Group Inc. is prohibited.

money to do proper development of the e-commerce shop. Transaction costs should also be low. The problem, however, is still to attract venture capital to ensure a stable financial environment.

*Feedback:* According to the theory, feedback is important in order that e-commerce entrepreneurs can assess if they are on the right track, their service is correct and their customers are happy and will visit their e-commerce shop again. It also means communicating the e-commerce vision inside and outside the organization.

*Management:* The problem that the authors stated in the theory is that top management should be fully committed to successful development of an e-commerce site. If top management does not commit them to it, then the project and e-commerce is doomed to failure. Roles for the team should be defined and responsibilities should be allocated.

*Marketing:* It is important to note the number of advertisements with URLs on them during visits to sport stadiums. Having a brilliant e-commerce site is no use without marketing and the authors have pointed out that successful e-commerce sites spend more money on advertising. It also pays to have a well-known brand in your stable, as this would ensure that people do visit the e-commerce shop. Changes should be marketed to partners, suppliers and customers.

*People (staff and customer):* The theory noted that well trained staff and customer care would ensure that the site is successful. Without customers, there would be no e-commerce and no survival. There should be input from end-users as well (from both sides). Individuals should be educated about new processes, systems and business practices.

*Alignment of goals:* The theory noted that the alignment of goals was a theme that successful organizations should keep in mind while conducting e-commerce. The responsibility rests with the e-commerce team to ensure that the e-commerce goal and the organization's goal should be aligned. The authors also noted that the e-commerce team should not be part of the IT department or should be separated once they had been picked.

*Items not suited for sale over the Internet:* the theory argued that many items are not suitable for sales over the Internet, like funerals, perfume, etc. Organizations

Copyright © 2003, Idea Group Inc. Copying or distributing in print or electronic forms without written permission of Idea Group Inc. is prohibited.

should spend some time building a brand name (it does not come overnight) and ensure that they also stock and advertise the items that will sell over the Internet.

*Parameters:* the parameters should be set for the e-commerce so that any variance to the original plan is noted immediately and that corrections are made on time. This would help to manage the vertical tension of centralization versus de-centralization and of separation versus integration of the e-commerce organization.

*Security:* Successful e-commerce sites ensure that their site is secure in terms of freedom from hackers and secondly that transactions cannot be copied. Many customers do not trust the Internet and a secure environment would ensure that everybody feels happy.

The purpose of developing the theory was to explain how e-commerce managers go about the business to assess its success factors. If a satisfactory explanation can be developed, it could provide a basis for evaluation of e-commerce management. This will allow an understanding as to the likely success of an e-commerce organization, and whether or not management complies with the factors expressed in the theory.

It can be deduced that the first four factors are important drivers (50%) that can ensure successful e-commerce start-ups. These four factors should be included when developing and running an e-commerce organization. The other factors are also important in ensuring a successful e-commerce. However, in the case of e-commerce, it is suggested that these could be of higher importance.

## Surveying the South African E-Commerce Practitioners

The framework developed (See Figure 2) was used as a basis to formulate a questionnaire. There were 50 organizations that were identified from a list of contacts supplied by the e-Business Centre of the Cape Technikon. The Web directory of TELKOM S.A. was also used extensively. A total of 120 questionnaires were e-mailed and posted out. This was distributed to 50 e-commerce organizations. Of these 27 were returned giving a 54% rate of return. The size of each of the participating organizations differed (from a giant financial organization to some smaller firms). A range of responses was obtained. Some organizations showed a high level of enthusiasm in the survey whereas indicated an indifference to the importance of the research. For example one big website run by a big shopping complex indicated that they could not complete the questionnaire and that they could not force their participants to complete the questionnaire.

Copyright © 2003, Idea Group Inc. Copying or distributing in print or electronic forms without written permission of Idea Group Inc. is prohibited.

## Empirical Findings

The factors listed in Figure 3, were listed alphabetically and these were required by the respondents to be ranked from 1 to 10 (1 being the most important and 10 the least important). On the second page, these factors were again listed alphabetically, and incorporated in statements. The respondents were required to mark a preference based on a Likert-type scale (1=strongly disagree to 7=strongly agree). The data was collated on an Excel spreadsheet and the average for the two columns were calculated. (See Table 3).

The rating in Figure 3 is the average of the total number of completed questionnaires. The ranking in the Column 2 was done based on the rating in column 1. Responses to the Likert scale were also ranked; they are presented in Figure 3. These ratings were entered onto a grid using SPSS for Windows and the Spearman correlation factors were calculated using the program (see Appendix).

## Discussion of Findings

The calculations of the Spearman correlation coefficient has shown that there was little correlation between the ranking of the critical success factors identified in the literature (overseas) and the ranking of their practicing South African counter-

*Figure 3: Rankings of Key Factors by Practitioners*

|    | Theme | Rating | Ranking | "Likert" rating | Content analysis ranking |
|----|-------|--------|---------|-----------------|--------------------------|
| 1  | Alignment of Goals | 4.5 | 7.5 | 4 | 7 |
| 2  | Capitalization | 5.6 | 3.5 | 9 | 2 |
| 3  | Feedback | 6.3 | 1 | 6.5 | 3 |
| 4  | Items suitable for selling | 4.3 | 9.5 | 1 | 8 |
| 5  | Management | 5.6 | 3.5 | 3 | 4 |
| 6  | Marketing | 4.5 | 9.5 | 6.5 | 5 |
| 7  | Parameters | 6 | 2 | 8 | 9 |
| 8  | People | 5.2 | 6 | 2 | 6 |
| 9  | Research and Development | 4.3 | 7.5 | 10 | 1 |
| 10 | Security | 5.3 | 5 | 5 | 10 |

Copyright © 2003, Idea Group Inc. Copying or distributing in print or electronic forms without written permission of Idea Group Inc. is prohibited.

parts (only –0.046). The Spearman factor for the correlation between the literature ranked factors and the ranking based on a Likert scale was negative (-0.405). This suggested that there was a better correlation between the literature-induced factors and the personal 'flashes of commercial insight' factors – although not the highest form of correlation.

The ranking between the Likert-type ranking and the literature was 0.218. This was also not high and possible reasons for this could be that the e-commerce revolution is in its infant stage in South Africa, and that local businesses are still figuring out how to handle a new business paradigm. There is a clear indication that more research is needed to support and lead the South African e-commerce organizations.

Some supporting documentation that was added by the respondents was used. They added that the factors in Figure 2 are all key factors. They are not sure that any e-commerce organization can get away without any of these elements in their business and they would not like to have a business plan without any of these aspects. Other important aspects they noted was that the customer has to perceive the site is secure, that information supplied is confidential and that it is a reputable company. Part of this would be to have a good product or service and an efficient fulfillment process (a key element according to some respondents). The biggest problem in South Africa is to get the completed product in the hands of the customer. One respondent noted that when a poor business and a great team come together, the poor business tends to win most of the time. They also noted that security depends on the needs of the customer. Some want the site be more efficient, but could not describe the parameters for this. Others wanted to ensure that the information is a factor, but customers will not know how if the information is complete or not.

Other significant comments made was that management needs to be part of the process and alignment is a key process. The plan will not get off the ground if management is not supporting it. The e-commerce site must be functional, simple and responsive. An idea could be to conduct some research in terms of function points and how these agree with customer's preferences. They all noted that people would come for the product; therefore, a slow, complicated site is not required. The advice was: "Get rid of all those graphics and allow your clients to get where they want to go." A satisfied customer is your best advertisement.

# CONCLUSION

Generally, a great deal of interest has been shown in the subject of e-commerce and its implementation. All the authors that this study had consulted were unanimous

Copyright © 2003, Idea Group Inc. Copying or distributing in print or electronic forms without written permission of Idea Group Inc. is prohibited.

in their opinion that many factors could affect e-commerce; these factors could vary from day-to-day. It appears from the pilot research that there is a trend towards combining the issues of formulation and implementation of e-commerce success factors. This is reflected in a generally higher level of awareness of organizations concerning the importance of applying success factors to form a successful e-commerce. The poor correlation between the conceptual factors derived form the literature and the survey, reinforce the fact that e-commerce activity is in a different level of development (infant stage) as compared to international counterparts. However, what is clear is that there are a number of lessons that can be learned from the experiences (success and failures) of others. South African e-commerce managers therefore should lean heavily on this in order to improve the effectiveness of their own Internet-based initiatives.

# REFERENCES

Abel, S. (2002) *Planning Business-to-Business Web sites: Avoiding the Common Mistakes*, Nims Associates, white paper.

Berelson, *cited* in Kroppendorf, K. (1980). *Content Analysis*, Sage Publications.

Business 2.0, Inc. (2002). *Boo! And the other 100 dumbest moments in e-business history*, white paper.

Chan, H., Lee, R., Dillon, T., & Chang, E. (2001). *E-commerce Fundamentals and Applications*. New York: Wiley.

Erwin, G.J. & Blewett, C. (1998). *Business Computing – An African Perspective*. Juta: Cape Town.

Harmon, P. (2002). *Reengineering your corporate e-business strategy*, Cutter Consortium, available at: http://www.cutter.com/itgroup/reports/reengineer.html.

Hayes, Linda G. (2002). *The (Questionable) Need for Speed*, Earthweb IT Management, available at: http://www.itmanagement.earthweb.com.

Highsmith, J., Yourdon, E., Bach, J., Becker, S.A., Berkemeyer, A.H., Bragg. T., McDonald, M., Scott, J. & Zou, B. (2001). *Best practices for e-project management*, Cutter Consortium, white paper.

Huberman, A.M. & Miles, M.B. in *Handbook of Qualitative Research*, Denzin N.K. & Lincoln Y.S., (eds.) (1994), London: Sage.

InterOPS Management Solutions (2001). *Protecting the brain: Ensuring e-business continuity for financial services*, white paper, available at: http://www.interops.com.

ITQuadrant. *10 Critical Factors behind Successful e-Business Implementations*, available at: http://www.ebizQ.net.

Copyright © 2003, Idea Group Inc. Copying or distributing in print or electronic forms without written permission of Idea Group Inc. is prohibited.

James, G. 1997. *IT Fiascoes and how to avoid them*, Datamation, available at: http://www.datamation.com/cio/11disas.html.

Johnson, J. (1995). CHAOS - the dollar drain of IT project failures, *Application Developments Trends*, January.

Kandiah, V., Mohd, S.H., Nasirin, S., Oz, E. & Sosik, J.J. (2000). *Information (IS) project failure: New evidence from Malaysia*. Working paper.

Kalakota, R. & Marcia, R. 1999. *E-Business: Roadmap for Success*. Reading: Addison-Wesley.

Lubbe, S.I. (1997). Ph.D. Thesis submitted to the University of the Witwatersrand, *The Assessment of the effectiveness of IT investment in South African organisations*.

Morell, J. & Swiecki, B.F. (2001). *E-readiness of the automative supply chain: Just how wired is the Supplier Sector?*, Erim, Available at: http://www.supplysolution.com.

Oakes, P. (2002). *The five myths of supply chain management*, TechRepublic, available at: http://www.techrepublic.com

O'Brien, J.A. (2002). *Management Information Systems (5th ed.)*. Boston: McGraw Hill.

Pickering, C. (2001). *Survey of e-Business and IT practices*, Systems Development, Inc. Available at: http://www.cutter.com/itgroup/reports/survey.html.

Wilder, C. (1998). *Myths and Realities*, Informationweek.com, available at: http://www.informationweek.com/712/12iumyt.htm.

Wilson, R.F. (1999). *Dot.Com Countdown: 7 Steps to e-Business Launch*, Web Marketing Today; Wilson Internet.

Wilson, R.F. (2001). *How to achieve Dot-Success in an era of Dot-Bust*, Web Marketing Today; Wilson Internet.

Copyright © 2003, Idea Group Inc. Copying or distributing in print or electronic forms without written permission of Idea Group Inc. is prohibited.

# APPENDIX

## Correlations

| | | | VAR00001 | VAR00002 | VAR00003 |
|---|---|---|---|---|---|
| Kendall's tau_b | VAR00 001 | Correlation Coefficient | 1.000 | -.046 | .163 |
| | | Sig. (2-tailed) | . | .856 | .525 |
| | | N | 10 | 10 | 10 |
| | VAR00 002 | Correlation Coefficient | -.046 | 1.000 | -.405 |
| | | Sig. (2-tailed) | .856 | . | .106 |
| | | N | 10 | 10 | 10 |
| | VAR00 003 | Correlation Coefficient | .163 | -.405 | 1.000 |
| | | Sig. (2-tailed) | .525 | .106 | . |
| | | N | 10 | 10 | 10 |
| Spearman's rho | VAR00 001 | Correlation Coefficient | 1.000 | -.012 | .218 |
| | | Sig. (2-tailed) | . | .973 | .546 |
| | | N | 10 | 10 | 10 |
| | VAR00 002 | Correlation Coefficient | -.012 | 1.000 | -.492 |
| | | Sig. (2-tailed) | .973 | . | .148 |
| | | N | 10 | 10 | 10 |
| | VAR00 003 | Correlation Coefficient | .218 | -.492 | 1.000 |
| | | Sig. (2-tailed) | .546 | .148 | . |
| | | N | 10 | 10 | 10 |

Copyright © 2003, Idea Group Inc. Copying or distributing in print or electronic forms without written permission of Idea Group Inc. is prohibited.

# Chapter X

# E-Learning is a Social Tool for E-Commerce at Tertiary Institutions

Marlon Parker
Cape Technikon, South Africa

## ABSTRACT

*This chapter sets out to inform the reader about the background of e-Learning. It starts with a brief introduction to the concept of e-Learning, presenting a compelling case why institutions implement e-Learning and describes the difference between technology-delivered e-Learning and technology-enhanced e-Learning. The discussion includes some advantages and disadvantages of technology-enhanced e-Learning and then examines some learner, facilitator and technology aspects of technology-enhanced e-Learning. It continues with a brief discussion on technology-enhanced e-Learning at the Cape Technikon.*

*This project further assesses the perceptions of learners using an online survey to address the issues and concerns that learners experienced with regard to technology-enhanced e-Learning. This chapter also inspects the efficiency usage of e-Learning tools within a technology-enhanced e-Learning environment and concludes with some significant findings of the survey, which includes the importance of computer literacy, interaction and communication in the technology-enhanced e-Learning environment.*

Copyright © 2003, Idea Group Inc. Copying or distributing in print or electronic forms without written permission of Idea Group Inc. is prohibited.

# INTRODUCTION

Tertiary education institutions aim to be recognized for social, knowledge and economic contributions in South Africa. There has also been an increase in the different uses of the Internet (e.g., online banking, online shopping, learning and teaching within tertiary educational institutions.) This increase has contributed to the electronic learning revolution and some South African tertiary institutions are making a technology-based paradigm shift for this reason.

The changes in delivery methods of Information Systems subjects are a suggestion of the technological changes in our society as a whole. There are some issues that are of concern to learners studying in a technology-enhanced e-learning environment. This study used an electronic survey to assess student perceptions of the technology-enhanced e-learning environment.

## Institution of Study

In 1920, HRH Prince Arthur of Connought laid the foundation stone of Longmarket Street Building of the then Cape Technical College. The establishment of the college followed more than ten years of representations by the community for the consolidation of the technical courses, which had been offered in various venues in town. In 1923, the official opening of Longmarket Street Building took place. The building soon proved to be too small; extensions were added in 1926 and 1949. The Cape Technikon is an institution that has taken on the e-learning initiative.

# BACKGROUND

Traditional delivery of learning methods generally required only the instructor, a textbook and support materials according to (Taylor, 2002). Traditional learning environments are defined in terms of time (the timing of instruction), place (the physical location of instruction) and space (collection of materials and resources available to the learner) (Piccoli et al., 2001).

The acceptance of the Internet and use of World Wide Web (WWW) technology in tertiary institutions have resulted in the surfacing of an endless amount of resources for students (Sheard et al., 2000). The Internet had an overwhelming impact on a number of industries (Evans & Wurster, 1997) and the growth in Internet usage created much interest in Web-based learning (Fong & Hui, 2002). (Tian, 2001; Fong & Hui, 2002) argued that students can access resources globally through the Internet to assist them in their learning and that it has become an attractive alternative to traditional modes of communication.

Copyright © 2003, Idea Group Inc. Copying or distributing in print or electronic forms without written permission of Idea Group Inc. is prohibited.

Piccoli et al. (2001) reported that technology courses were among the early offerings on the web and motivation for e-learning in tertiary institutions, particularly in Information Systems education, arises from the search for efficient delivery vehicle for course content.

# LITERATURE STUDY

## Introduction

Information technology is important to modern societies and their educational systems according to Carter & Boyle (2002). Hansen et al. (1999) reported that the increasing acceptance and availability of access to the Web, various web-based teaching initiatives, are either being developed or being adopted by educational institutions.

Serdiukov (2001) therefore suggested three new models of learning on the web, supporting Carter & Boyle. The models are:

- *Teacher-Student*
  This is a traditional model for education with a dyad between student and teacher.

- *Teacher-Computer-Student*
  The Teacher-Student model has been transformed into the Teacher-Computer-Student model, which has more advantages because it allows for Student-Computer, Student-Teacher and Teacher-Computer interaction. Technology qualitatively changes the relationship between people and knowledge according to Eisenstadt (1995).

- *Computer-Student*
  This model terminates live human presence, which mean there is no Teacher-Student contact time.

There is an interaction of teaching, learning and enabling elements on the World Wide Web that facilitate cognitive change in learners (Rogers, 2001). However, no matter what kind of learning environment an institution uses, these institutions will be sensitive to these elements. These elements are elements of convenience, interactivity, flexibility, access and availability of suitable and supportive technologies, and quality assurance (Rogers, 2001).

Copyright © 2003, Idea Group Inc. Copying or distributing in print or electronic forms without written permission of Idea Group Inc. is prohibited.

Chin et al. (2000) reported that the online learning environment is potentially an arena in which new practices and new relationships can make significant contributions towards learning. However, Jones et al. (2000) identified the Internet as an ideal vehicle for delivering of course material and providing students with flexibility, reliability and freedom of choice. Slay (1997) agreed that there is a current trend towards the delivery of courseware via the World Wide Web. Wells (1999) expanded that the Internet is primarily a delivery vehicle for information, which assists with the facilitation of the teaching and learning process.

The above-mentioned authors agreed that the Internet has an important role to play within education. The Internet being used, as only a delivery vehicle might not be the most effective usage of the tool, as the literature will discuss in further sections, but as an interactive tool, it will contribute towards the learning process as reported by Rogers (2001).

## Education and Technology

Zepke (1998) argued that technology enthusiasts declared a marriage between the Internet and education. Robson (2000) also agreed that visionary educators have seen computers as tools to assist in educating students. Schulze (2000) added that the quality of higher learning should be improved and teaching-learning programs should be technology-based.

Rogers (2001) suggested that educators are concerned with the value of the education in distance learning environments. Learners expect the learning outcome to be of high quality, convenient, personal and interactive according to Rogers (2001). Sutliff and Baldwin (2001) agreed that if material is presented without personalizing it and without student interaction, students would experience difficulty understanding the course content.

Sherson (1999) noted that the advances of the digital revolution would encourage teachers to make effective use of digital resources and support them in the relationship between information and education. Jones et al. (2000) agreed with Sherson that the Internet is an ideal delivery vehicle of course material and providing students with flexibility, reliability and freedom of choice.

According to Serdiukov (2001), modern Educational Technology (ET) is based on computer and telecommunication technologies. Besides the huge databases that are available on the World Wide Web (WWW or Web), it also allows for the concatenation of educational forces. Education is a complete, dynamic system that develops and transforms from one condition to another under the influence of external factors further as suggested by Serdiukov (2001):

Copyright © 2003, Idea Group Inc. Copying or distributing in print or electronic forms without written permission of Idea Group Inc. is prohibited.

*"One particular factor is the growing sophistication of human activity under the impact of the new Information Technologies and the resulting need for continuous, lifelong professional development. Another factor is the internal urge within the system of education for improvement to survive in the period of dramatic external changes"*

Marold (2002) argued that the personal computer is the "universal tool" of the 21st century. Ndahi (1999) also noted that technology influences on learning and teaching are more evident within educational institutions. However, some educators still continue to rely on traditional content delivery methods and refuse to accept the technologies available. The reliance on only traditional delivery methods might therefore have a negative impact on tertiary institutions, which have the responsibility to reach students at any time and any place. Technological innovations in the telecommunications industry have improved teacher-student interaction (Ndahi, 1999).

Slay and Jones et al. mentioned that educators are evaluating the use of the Internet in creating an effective learning environment. Tertiary institutions could maintain effective, current and high quality education if Internet technologies are embraced (Slay, 1997; Jones et al., 2000).

The web medium has the potential for group and collaborative work with staff and students both locally and between institutions and it is natural for education institutions to have adapted personal computers to its educational needs (Marold, 2002).

Cuban (1993) supported the above mentioned and added some goals for the introduction of computers in education:

- Keep the education system at the forefront of technological development and students' skills up-to-date with those expected in the workforce.
- Increase efficiency and productivity in teaching and learning.
- Enable more self-directed learning, with students as active learners assisted by teachers to construct their own understanding.

The above literature indicates how some authors recommended that technology and the Internet be used within tertiary institutions. However the costs involved, with these technologies, has become a barrier with the full realization of usefulness of technology within education. Technology also needs to enhance the learning process rather than becoming another delivery medium. Quality of learning should also be assured by the introduction of technology within tertiary institutions.

Copyright © 2003, Idea Group Inc. Copying or distributing in print or electronic forms without written permission of Idea Group Inc. is prohibited.

## The Virtual Classroom

Tiffin & Rajasingham (1995) first used the term "virtual class" to refer to the learning process enabled solely by telecommunications and distinguishes the concept of the virtual class from the "virtual classroom" proposed by Hiltz (1996) as "...it suggests that the place a virtual class is held is an electronic simulation of conventional classroom" and further described it as the use of computer-mediated communications "...to create electronic analogue of the communications forms that usually occur in a classroom including discussion as well as lectures and tests".

Uys (1999) agreed with Tiffin & Rajasingham, that the virtual class could be described as the process that occurs when teacher, learner, problem and knowledge are joined solely through communication and information technologies for the purpose of learning and teaching. (Piccoli et al. 2001) referred to the virtual class as virtual learning environments (VLEs) and (Wilson, 1996) defined it as "computer-based environments, allowing interactions and encounters with other participants".

Uys (1999) noted that the virtual class is an educational experience of real people in a virtual dimension and that in the virtual class the teaching and learning is performed without the movement of physical objects (e.g., getting students and lecturers into a physical venue). Dede (1996) argued that the virtual classroom has a wider spectrum of peers with whom learners can collaborate than any local region can offer and a broader range of teachers and mentors than any single educational institution can afford.

The virtual class as defined by Tiffin & Rajasingham (1995) and Uys (1999) supports the Computer-Student model suggested by Serduikov (2001). The virtual class can be described as a technology-delivered e-learning environment.

## Electronic Learning

### The Internet in Tertiary Institutions

There is a steady increase in the use of the Internet and the World Wide Web for tertiary education according to Chin et al. (2000). Convenience and flexibility of web-based delivery have attracted many students who are unable to attend traditional classrooms.

Sherson (1999) argued that there is a growing need for education institutions to digitize their content and activities. Technology should also be used for developing curriculum and supporting the teacher. The digital revolution is in progress and teachers will need to be encouraged make effective use of digital resources (Sherson, 1999).

Copyright © 2003, Idea Group Inc. Copying or distributing in print or electronic forms without written permission of Idea Group Inc. is prohibited.

## Types of E-Learning

Jackson (2001) divided electronic learning into technology-delivered e-learning and technology-enhanced e-learning.

- Technology-delivered e-learning

Technology-delivered e-learning is where the learner audience is never in physical proximity to the instructor and may be delivered via a blend of asynchronous and synchronous technologies. It is also known as "Distance Education," "Distributed Education" or "Distance Learning."

- Technology-enhanced e-learning

Technology-*Enhanced* e-learning is where the learner audience has the opportunity to meet face-to-face with the instructor. It is a supplement to traditional, on campus learning and replaces materials previously delivered to students as "shrink-wrap" courses. Typically includes online syllabi, bibliographies (often hyperlinked), and faculty backgrounds and instructor-led sessions are live, face-to-face in traditional classrooms. The typical technology-enhanced e-learning asynchronous technologies are implemented through either a web editor or an asynchronous course management system. Taylor (2002) agreed with Jackson and stated that e-learning can be used effectively in several different forms.

## The Aims and Tool Usage of E-Learning

It can be used as a stand-alone asynchronous program, or as a synchronous class where all the students are online at the same time, or as an add-on to traditional classroom presentations. With regard to add-on to classroom presentations, e-learning could be part of learning delivery systems for most courses; however, not every course is suited for complete online presentation. In a stand-alone asynchronous program, students can access course materials and lessons at any time and at any place. Communication between learner and facilitator is via electronic means and tools such as e-mail and bulletin boards are used. In a synchronous classroom, students access the lessons at the same time and communication between students and facilitator are via Internet Relay Chat (IRC) or I Seek You (ICQ) (Taylor, 2002). Alexander (2001) further argued that using technology in both classroom and distance learning would improve the quality of learning, the access to education and training, reduce the costs of education and improve the cost-effectiveness of education.

Copyright © 2003, Idea Group Inc. Copying or distributing in print or electronic forms without written permission of Idea Group Inc. is prohibited.

The following aims to incorporate technology into an established program were suggested by O' Keefe & McGrath (2000):

- To enhance students' development of a range of generic skills.
- To add to the inventory of teaching and learning methods the use of technology.
- Broaden the content of the course.
- Utilise conference formats and encourage students to work as peer groups.
- Mass distribution of information avoiding repetitive questions.
- To provide a flexible learning environment.
- Greater time flexibility for learners and facilitators

## Advantages of E-Learning

*Advantages for the Learner*

Delivery of Information Technology education is advantageous in several ways (Marold, 2002). He noted that there is a movement from text to multimedia that delivers a richer content to the students. The delivery organization is non-linear, meaning that the content can be hyperlinked in a link by association fashion. This allows the learner to access this richer content in any order (flexibility) and at a time that is appropriate. Accessibility is universal and learners can access content when required. Changes in Information Systems development, hardware and software releases, project progress and course content are reflected immediately. There is an ease of content and schedules updates.

The other advantage according to Marold (2002) is that the multimedia presentations are advancement over the written notes on the blackboard and handwritten transparencies. Being able to interact and be actively engaged in the learning process is a real advantage over simply observing the teacher's demonstration. The access to course content is also controllable by the facilitator. If there is no interaction by the learner then no learning can take place in a technology-enhanced e-learning environment. The learner is therefore essentially responsible for the success of the learning process. There are many resources available to the learner to which they have direct access (Jackson, 2001).

*Advantages for the Faculty*

Taylor (2002) discussed benefits of technology-enhanced e-learning from a tertiary institution faculty perspective. E-learning offers the instructor the ability to greatly enhance his or her presentation by adding multimedia features. Students, who feel intimidated by others in a traditional class, will participate if they can do so using the keyboard. There is an overall learning curve with regards to technology

Copyright © 2003, Idea Group Inc. Copying or distributing in print or electronic forms without written permission of Idea Group Inc. is prohibited.

for students as well as facilitators. E-learning improves the distribution of course materials to students. Course materials can be distributed immediately and the lengthy turnaround time is eliminated. Online communication through e-learning also encourages student inter-communication, which gives the students feedback from their peers as well their facilitators. There are quite a number of classroom management software programs available, e.g., the Cape Technikon is using WebCT. Students can receive immediate and meaningful feedback from interactive questionnaires and quizzes. Updating course materials can be done at ease and speed.

Hansen, et al. (1999) supported Marold, Jackson, Taylor and O'Keefe & McGrath's statements and identified the following promising features of e-learning which makes it attractive:

- Global accessibility
- Inherently multi-media oriented
- Inherently interactive
- Allows team and shared work
- Ability to upload submitted material
- Team delivery
- Amalgamation to existing global resources
- Integration to an organization's information systems
- Asynchronous and synchronous communication

E-learning also provides clear accountability for all participants in the learning process (Gunasekaran et al., 2002).

## Disadvantages of E-Learning

Jackson (2001) suggested that the bandwidth and browser limitations might restrict instructional methodologies and that limited bandwidth means slower performance. Time required for downloading applications due to the limited bandwidth are another negative of technology-enhanced e-learning. Student assessment and feedback is limited and it cannot easily be reordered from its original state (Marold, 2002). Learning process takes place individually and self-paced, which means that group discovery that characterizes the ideal classroom situation is lost and that the emphasis the teacher places on course content might differ from the student's individual choice. There are also fixed costs involved with web-based course delivery.

The above-mentioned authors reported that the advantages needs to be weighed against the disadvantages and the type of e-learning concept to follow are important factors with regard to the implementation of e-learning. Ways to satisfy

Copyright © 2003, Idea Group Inc. Copying or distributing in print or electronic forms without written permission of Idea Group Inc. is prohibited.

learners and keep them engaged in the online learning process need to be discovered as society move forward into the digital age (Hill, 2001).

## Three Pillars of Technology-Enhanced E-Learning

Serdiukov (2001) suggested a Teacher-Computer-Student model, which consists of several components of technology-enhanced e-learning (See Figure 1).

*The Learner*

Jones et al. (2000) suggested that proficiency in computer-based multimedia is an important skill to students who use the Internet on a fairly regular basis. (Piccoli *et al.* 2001) divided e-learning into the following dimensions:

- Human dimension
- Technology dimension

The learner and facilitator form part of the human dimension and technology form part of the design dimension. Students have found themselves in a comfort zone within the classroom and are the primary participants in any learning environment. E-learning has noticeably increased the student use of technology and has shifted control and responsibility to the learners. E-learning allows a flexible dimension in terms of any time, place and space, which are very attractive to students who have difficulty attending lectures. All the participants, learners and facilitators, interact extensively with ICT. The learners who are comfortable with technology and have a positive attitude toward it are more likely to succeed within an e-learning environment.

Rossiter (1999) agreed that computer literacy is being added to the list of generic capabilities, as an attribute needed by all students who effectively wants to

*Figure 1: Teacher-Computer-Student Model of Learning*

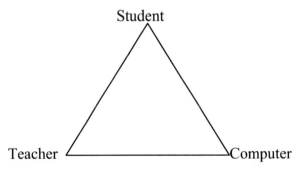

*Source: Serdiukov, 2001*

Copyright © 2003, Idea Group Inc. Copying or distributing in print or electronic forms without written permission of Idea Group Inc. is prohibited.

participate in further studies. Technology literacy is one of the foundation blocks of technology-enhanced e-learning. When students have inadequate technology skills, the educator has to either refer students to generic skills course or attempt to teach the required technology skills themselves. The lack of technological skills among students' hampers the learning opportunities offered by teachers and acts as a barrier to effective learning for students. Nasseh (1996) stated that:

> *"The technological skills for utilization and operation of computer applications and tools are very essential for teachers and students who participate in computer-based distance education."*

Uys (1999) added that both the learner and facilitator need a sophisticated level of computer literacy and use. While the online facilitator has the necessary technological support within the tertiary educational institute the students often may not have the same advantage. Technological literacy is one of the foundation blocks of modern-day learning (Uys, 1999). Success in computer-based learning is often seen as requiring additional skills of students in taking greater responsibility of their own learning (Akerlind & Trevitt, 1999).

Technological literacy is one of the foundation blocks of modern-day learning and technology education must prepare students to understand, control and use technology (Boser et al., 1998). Students tend to focus on the web-based delivery as an aid to collect information rather than using the system in a more interactive way. Students appear to be demanding more technology in tertiary education; they are not demanding technology as a substitute for face-to-face teaching. Students still require support from their lecturers according to (O'Keefe & McGrath, 2000).

Laurillard (1993) established that interaction between teacher and learner and feedback from the teacher is a core element of learning, which suggests that successful communication is interactive, adaptive and reflective. Passerini and Granger (1999) agreed that learning is a product of interactions in the virtual classroom. Passerini and Granger (1999) further pointed out that in an online environment there are several channels of communication (student-content, student-to-student, student-to-instructor, student-to-other-hypermedia content and student-to-other-instructors). Students in the virtual classroom, through an extended group of online facilitators, can therefore be challenging and also knowledgeable than the conventional student (Uys, 1999).

Retails and Avgeriou (2002) suggested the following roles of the learner in their web-based Instructional Systems model:

- Attends lectures
- Navigates freely within the learning resources
- Explores information resources

Copyright © 2003, Idea Group Inc. Copying or distributing in print or electronic forms without written permission of Idea Group Inc. is prohibited.

- Asks questions
- Collaborates with other classmates in team projects
- Seeks feedback on discussion topics
- Interacts with other learners either face-to-face or via e-mail

Technology skills are one of the important aspects that learners need to have to ensure a high quality of learning within an e-learning environment (Jones et al., 2000).

## The Facilitator

Laurillard (1993) pointed out that the teacher is an important mediator in the process of constructive academic learning. O'Keefe and McGrath (2000) argued that it is problematic that not all lecturers have the knowledge and skill of the use of technology. Pennel (1996) agreed and suggested that teaching will increasingly make use of Internet technologies allowing students flexible access. Educators have to learn many new skills in order to maintain their competency as educators, argued Pennel (1996).

Sherson (1996) further described the role of the facilitator as being essential and be able to identify the areas where the learner needs motivation and be able to provide structured and incidental resources, instruction, direction, feedback and support to assist the learning process. These are essential in any form of programme delivery including the use of the World Wide Web. Dede (1996) stated that the facilitator needs to teach the learner how to learn and construct their own knowledge by making sense of *"... massive, incomplete and inconsistent information sources"*.

O'Keefe and McGrath (2000) identified the following three areas of increased workload for lecturers using online delivery:

- Additional time to learn the online system's pedagogical options and technical features.
- Time to undertake subject planning, design and material development and set up on the system.
- Time taken in on-going monitoring, fine-tuning, technical trouble-shooting and the increased asynchronous communication.

The facilitators need to alter their role in this new learning environment by understanding the capabilities of technology to be used and consider the role of technology in both delivery and curricula development of material (O'Keefe & McGrath, 2000). Retails and Avgeriou (2002) suggested the following roles of the facilitator in their Web-Based Instructional Systems model:

Copyright © 2003, Idea Group Inc. Copying or distributing in print or electronic forms without written permission of Idea Group Inc. is prohibited.

- Organises content into learning resources
- Gives lectures either face-to-face or by distance
- Displays and updates information about the course
- Brings up discussion topics
- Provides corrective feedback
- Advises and tutors students
- Assesses the students
- Monitors students' progress
- Creates and manages project teams

(Taylor 2002) argued that facilitators are faced with steep learning curves and that the quality of their online lessons depends heavily on their knowledge of the required technologies and software. The e-learning environment that is created by facilitators should be one in which learning can and will take place. (Webster & Hackley, 1997) agreed and pointed out that an instructor's positive attitude toward technology, interactive teaching style and control over technology contributed to some of the successes of effective learning.

Facilitators need the answers to some fundamental questions according to (Taylor, 2002):

- What are the advantages for converting from a traditional to a virtual class?
- Are there any benefits for the students?
- Are there benefits for the tertiary institution or organization?
- Are there any personal benefits?

The facilitator needs to identify potential students and address their learning needs in an online environment (Simpson, 2000). The prevailing emphasis on developing deep approaches to learning, facilitators may frustrate legitimate surface motivation and so inhibit student learning according to Zepke (1998). Zepke (1998) further stated that lecturers should be able to facilitate students learning by communicating and empathizing with them, structuring knowledge and arranging a reasonable workload.

(Piccoli et al., 2001) pointed out that the facilitator needs to be flexible because of the high time and energy demands. Courses where the facilitator is unable to meet these demands will most likely not be effective. Facilitator attitudes and actions can have an important influence on students' perceptions of e-learning.

## E-Learning Technologies

Wells (1999) argued that the delivery of on-line instruction is heavily dependent on network technologies. (Piccoli et al., 2001) agreed and further pointed out

Copyright © 2003, Idea Group Inc. Copying or distributing in print or electronic forms without written permission of Idea Group Inc. is prohibited.

that technology quality and reliability, accessibility is important determinants within the e-learning environment. Serdiukov (2001) divided technologies into two parts. The first being computer technology, which offers computer-based courses, computerized tests, word processors, graphics software, spreadsheets, databases and presentation software to the learning process. The second part is telecommunications technology, which offers distance courses, distributed educational resources, e-mail, video-conferencing, bulletin boards, whiteboards and chat rooms.

Roderic (2002) agreed and suggested some technologies vital to the technology-enhanced e-learning process. Text is likely the most common used media types on the web. Audio or Video also primary multimedia components assists bringing about richer course content. Data in the form of quantitative or qualitative aspects can be a very important aspect of online learning. E-mail and instant messaging are technologies that are useful within the online technologies architecture. Bulletin boards or discussion groups and newsgroups are forums, which contains lists of messages posted by users. Chat is another method of interaction between students and facilitators or students and students. Shared applications and remote control software are the technologies that allow users to share applications with selected users. Simulations are the technology that attempts to model the environment where learning takes place. Peer-to-Peer or File transfer protocol are technologies that are used for sharing of files. The same e-learning technologies can be used to support different learning models according to (Piccoli et al., 2001).

E-learning technologies will increase as a channel of communication between learner and teacher (Pennel, 1996). Tools such as email, file transfer protocol, computer conferencing and web browsing demand a serious learning effort from their users. Not only students as learners but also facilitators as administrators. This extra learning effort acts as a barrier between learner and facilitator in the learning process according to Pennel (1996). A partial list of desired communications includes (see Table 1):

Application of e-learning technologies for the sake of going online will not improve the learning process. Educators need to be aware that technology should be used to enhance the learning process, rather than replicating the current practice (O'Keefe & McGrath, 2000).

## E-Learning at the Cape Technikon

The author was involved with the implementation of e-learning in the faculty of Business Informatics at Cape Technikon. The technology-enhanced e-learning approach was adopted and the course management software system being used is WebCT. WebCT, at Cape Technikon, is an adjunct to the traditional lectures. It

Copyright © 2003, Idea Group Inc. Copying or distributing in print or electronic forms without written permission of Idea Group Inc. is prohibited.

*Table 1: Desired Communications (Source: Pennel, 1996)*

| No. | Communication | Typical tools |
|---|---|---|
| 1. | Free-form text messages between lecturer and individual students and vice-versa | Email software, Eudora |
| 2. | Messages from lecturer to entire class | Mailing list software |
| 3. | Delivery of learning materials and assignments to student when, where and as they need them | Telnet, Fetch or Gopher session, or a Web page with links to download files to the student's own machine |
| 4. | Response by student to short structured questions to allow lecturers and students to assess learning achieved | Forms-based Web pages backed with CGI scripting to extract results |
| 5. | Text-based commentary between students about the learning resources | IRC Chat session or computer conference |
| 6. | Transfer of image files between students for discussion, together with manipulation of such images visible to many students concurrently | Whiteboard facility included in many forms of desktop video conferencing |
| 7. | Submission of assignments by students electronically, together with return of assignments to students by the lecturer | e-mail with attached documents |
| 8. | Synchronous audio communication between a student or students and the lecturer | Audio-conferencing equipment. Party-line call or Internet Phone |
| 9. | Asynchronous audio communication between relevant parties | Answering machine or audio download from a Web page |
| 10. | Face-to-face visual communication between a student or students and the lecturer | Video-conference, desktop video-conference |

is used in and out of class as the facilitator decides the learning structure of the class. Cape Technikon currently has 196 registered courses using WebCT.

WebCT provides a secured environment for the placement of course materials on the Web. WebCT also uses Web browsers as the interface to produce the courses and for the course-building environment. It also has a variety of tools and features that can be added to a course. Students at Cape Technikon can access their WebCT course materials using a Web browser (Netscape and Internet Explorer are used at Cape Technikon) from any computer connected to the Internet or the technikon's Intranet. Students, too, can place assignments and other materials in WebCT in courses for which they are registered. It also provides course management tools, for facilitators, for grading, tracking student interaction and monitoring class progress.

## Summary of the Literature Study

The relationship between the Internet and education are inseparable according to Zepke (1998) and Rogers (2001) pointed out that the interaction of teaching,

Copyright © 2003, Idea Group Inc. Copying or distributing in print or electronic forms without written permission of Idea Group Inc. is prohibited.

learning and the Internet facilitates a cognitive change in learners. Some authors (Sheard et al., 2000; Tian, 2001; Fong & Hui, 2002; Sherson, 1999; Jones et al., 2000) all agreed that the Internet usage provide students with flexibility, reliability and freedom of choice, has increased within tertiary institutions and are being used as an ideal delivery vehicle within a web-based learning environment.

This has led to e-learning that Jackson (2001) divided into technology-delivered and technology-enhanced e-learning. Piccoli et al. (2001) categorized the learner, facilitator and e-learning technologies into the human and design dimensions respectively. Serdiukov (2001) suggested a Teacher-Computer-Student model, which consists of the learner, facilitator and technology.

Some authors (Jones et al., 2000; Rossister, 1999; Nasseh, 1996; Uys, 1999; Boser et al., 1998) all agreed that computer literacy is an important factor in the e-learning process. Interaction and communication were the other important factors that should also be considered when developing an e-learning course (Laurillard, 1993; Sherson, 1996; O'Keefe & McGrath, 2000). The facilitator therefore needs to understand and use appropriate e-learning technologies to ensure the quality of the learning experience.

*The Basis for Research*

A pilot study was done on the topic of issues and concerns of learners using technology-enhanced e-learning. The main objectives of this pilot study was:
- To assess the effective use of e-learning tools used at the Cape Technikon.
- To determine the perceptions of first year Information Students at the Cape Technikon.

## The Research Method

There are currently over 500 first-year students registered for Information Systems with the faculty of Business Informatics at the Cape Technikon. An online questionnaire as a survey method was used for the period August 5-9, 2002. The survey targeted first year Information Systems students at the Cape Technikon and 165 responses were achieved. The questionnaire instrument consisted of the following parts:
- Demographic information about the respondents
- 29 statements, using a five-point Likert-scale (Cooper & Emory, 1991), of which 12 statements were based on e-learning technologies.

All 29 statements were equally weighted. The response to each statement of each respondent was totaled to obtain total points scored for each respondent. An item analysis, for each respondent, was carried out to validate the degree of

Copyright © 2003, Idea Group Inc. Copying or distributing in print or electronic forms without written permission of Idea Group Inc. is prohibited.

discrimination for each statement. Only statements that were good discriminators (based on $t$-test) were used in further analysis (Cooper & Emory, 1991). Only those statements with a $t$-value of more than and equal to 1.75 were used (Edwards as cited by Cooper and Emory, 1991). Statements 8, 15 and 17 were excluded from further study based on that analysis.

## Results of the Study

*Student Demographics*
Of the 165 respondents, 108 were male (65.5%). The majority of the students were Information Technology students (77%) and about 66 percent of the respondents were under the age of 21. Only 35 of the respondents had no computer experience before taking Information Systems as a subject. The majority of the students said that they use the Internet more than 20 times a month (60%).

## Findings

More than 60 percent of the respondents agreed or strongly agreed that online lectures allow them to look up needed information more conveniently. Majority of the respondents agreed that learning is more fun with online lectures (47%). Only 13 respondents disagreed or strongly disagreed that they can easily reference any section of the lecture in an online environment. 117 of the respondents agreed or strongly agreed that communication between student and facilitator is crucial in the e-learning process. The same amount of respondents agreed or strongly agreed that interaction among students is crucial to the e-learning process. Most of the respondents agreed or strongly agreed that they received sufficient instruction on the use of the e-learning programs (74%). Only 16 of the respondents disagreed or strongly disagreed that their computer literacy was adequate for performing the functions required of the e-learning tools. A majority of the students agreed or strongly agreed that using e-learning as a platform for Information Systems has increased their course workload (36%). Most of the respondents agreed or strongly agreed that e-learning helped them learn the subject content more quickly (60%). Only about 10 percent of the respondents disagreed or strongly disagreed that e-learning has helped them develop new computing skills. About half of the respondents agreed or strongly agreed that they prefer having the subject material online than using a textbook.

E-learning technology statements retained for further analysis after item analysis was ranked in descending order of mean score value. The three e-learning tools identified as most effectively being used (high mean values) were assignment tool with a mean of 4.14, quiz tool with a mean of 4.12 and downloads tool with

Copyright © 2003, Idea Group Inc. Copying or distributing in print or electronic forms without written permission of Idea Group Inc. is prohibited.

a mean of 4.02 (see Figure 2). The author interprets these tools as being important to the respondents of this survey.

The three e-learning tools identified as least effectively being used (low mean values) were the chat room with a mean of 2.72, bulletin board with a mean of 3.04 and e-mail with a mean of 3.21 (see Figure 3). The remainder of the e-learning tools is ranked in Table 2.

*Figure 2: Most Effectively Used E-Learning Tools*

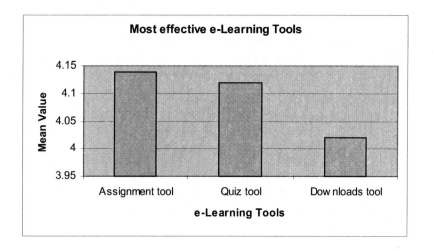

*Figure 3: Least Effectively Used E-Learning Tools*

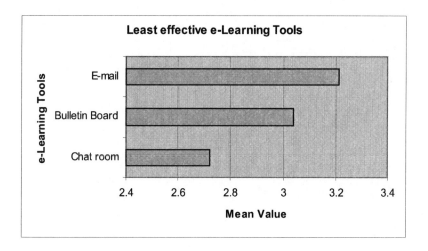

Copyright © 2003, Idea Group Inc. Copying or distributing in print or electronic forms without written permission of Idea Group Inc. is prohibited.

## Discussion of Findings

This survey has entertained the idea that students perceive interaction and communication to be factors to be taken in consideration within the e-learning process. Therefore, within a technology-enhanced e-learning environment the communication technologies used, need to be evaluated by the facilitator and the following questions should be asked:

- Are the learners as well as the facilitator skilled in using communication tools such as e-mail, chat rooms, bulletin boards, etc.?
- Is the subject facilitator utilizing the appropriate communication tools?
- Does students have problems accessing these communication tools?

The findings supported the following perceptions:

- An increase in course workload which supports statements by authors (Piccoli *et al.*, 2001) and (Rossiter, 1999).
- Learners find learning to be more fun and interesting with e-learning and information are easily accessible and easier memorized. This supports the author Alexander's (2001) argument that technology-enhanced e-learning would improve the quality of learning.
- There has been an acceptance of the Internet and an increase in World Wide Web (WWW) technology usage in tertiary institutions. (Sheard et al., 2000). A majority of the respondents in this study used the Internet 20 or more times per month.

*Table 2: Remainder of E-Learning Tools Ranked by Mean Value*

| e-Learning Tools | Mean Value |
|------------------|------------|
| Home page tool   | 3.78       |
| My Grades tool   | 3.63       |
| Self-test tool   | 3.48       |
| Search tool      | 3.47       |
| Discussion tool  | 3.38       |
| My progress tool | 3.38       |

Copyright © 2003, Idea Group Inc. Copying or distributing in print or electronic forms without written permission of Idea Group Inc. is prohibited.

Respondents to the questionnaire indicated the most effectively used technology-enhanced e-learning tools were the assignment and quiz tools. The following common attributes between the most effectively used e-learning tools indicated by respondents are:

- Assessment
- Feedback
- Direct contribution by facilitator
- Passive interaction

The technology-enhanced e-learning tools ranked as being the least effectively used were chat rooms, bulletin boards and e-mail. Taylor (2002) pointed out that the asynchronous and synchronous tools mostly used in technology-delivered e-learning are chat rooms, bulletin boards and e-mail. The results of this study clearly show and support Taylor that the tools being effectively used within a *technology-enhanced* e-learning environment will differ from those within a *technology-delivered* e-learning environment.

A distinct attribute of chat rooms, which were the least effectively being used from all e-learning tools, are interactive interaction whereas the assignment tool, which were the most effectively being used from all e-learning tools, have passive interaction.

# CONCLUSION

Most first-year information technology and financial Information Systems students perceive communication and interaction among themselves and facilitators to be of importance. Although e-learning contribute to the increase of their workload, they perceive learning to be more fun and of a better quality within the technology-enhanced e-learning environment.

Students perceive communication tools such as e-mail, chat rooms and bulletin boards as not being used effectively within technology-enhanced e-learning environments. However this study has highlighted that communication and interaction are important to the learning process, therefore the appropriate communication tools should be used more effectively within a technology-enhanced environment. This study has also highlighted those passive e-learning technologies (e.g., assignment tools, quizzes and download tools) are being more effectively used than interactive e-learning technologies (e.g., chat rooms, e-mail and bulletin boards).

Further work is needed to provide a technology-enhanced e-learning environment that will ensure quality and high performance learning and teaching experiences for both learners and facilitators. To insure that technology-enhanced e-

Copyright © 2003, Idea Group Inc. Copying or distributing in print or electronic forms without written permission of Idea Group Inc. is prohibited.

learning is being used, implemented and structured effectively. A web-based course development framework for facilitators or all educators who intend using e-learning as a teaching medium. A study on technologies used within the e-learning environment in correlation with multidiscipline subject matter:

- An investigation of e-learning tools in correlation with multidiscipline subject matter.
- A further study with a larger sample size as well as learners of other South African institutions using technology-enhanced e-learning as a delivery method could be useful.
- A study on facilitators' perceptions on the effectiveness of e-learning tools used in their course.

Some limitations of this pilot study were:

- Only Information Technology and Information Systems learners were surveyed.
- Learners from other subject areas such as engineering and management might have different perceptions on effectiveness of e-learning tools.
- The learner, facilitator and technology are not the only pillars in the e-learning equation. Some of the pillars that were not part of this pilot study are content, interaction, performance, satisfaction, and learning model (Piccoli et al., 2001).

# REFERENCES

Akerlind, G. & Trevitt, A.C. (1999). Enhancing learning through technology: When students resist the change. *Innovations in Education and Training International, 36* (2), 96-105.

Alexander, S. (2001). E-learning developments and experiences. *Education + Training, 43* (4/5), 240-248.

Boser, R., Palmer, J. & Daugherty, M. (1998). Students' attitudes toward technology in selected technology education programs. *Journal of Technology Education, 10* (1), 6-19.

Carter, J. & Boyle, R. (2002). Teaching delivery issues – Lessons from computer science. *Journal of Information Technology. 1* (2), 77–89.

Chin, K., Chang, V., & Bauer, C. (2000). The use of web-based learning in culturally diverse learning environments. The sixth Australian World Wide Web Conference. Retrieved on 17 April 2002 from the World Wide Web: http://ausweb.scu.edu.au/aw2k/papers/chin/paper.html.

Copyright © 2003, Idea Group Inc. Copying or distributing in print or electronic forms without written permission of Idea Group Inc. is prohibited.

Cooper, D. & Emory, C. (1991). *Business Research Methods.* Richard D. Irwin, Inc., USA.

Cuban, L. (1993). Computers meet classroom: Classroom wins. *Teachers College Record, 95* (2), 185-210.

Dede, C. (1996). The evolution of learning devices: smart objects, information infrastructures, and shared synthetic environments. *The Future of Networking Technologies for Learning.* [Online]. Retrieved on 12 April 2002 from the World Wide Web: www.ed.gov/Technology/Futures/dede.html.

Eisenstadt, M. (1995). Overt strategy for global learning. *Times Higher Education Supplement.* Multimedia section: April, 6-8.

Evans. P. & Wurster, T. (1997). Strategy and new economics of information. *Harvard Business Review, 75* (5), 70 – 93.

Fong, A & Hui, S. (2002). An end-to-end solution for Internet lecture delivery. *Campus-Wide Information Systems, 19* (2), 45-51.

Gunasekaran, A., McNeil, R. & Shaul, D. (2002). E-learning: Research and applications. *Industrial and Commercial Training. 34* (2), 44-53.

Hansen, S., Deshpande, Y. & Murugesan, S. (1999). *Adoption of Web Delivery by Staff in Education Institutions. Issues, Stratagems and a Pilot Study.* The Fifth Australian World Wide Web Conference, Southern Cross University. [Online]. Retrieved on 23 May 2002 from the World Wide Web: www.ausweb.scu.edu.au/aw99/papers/hansen/paper.html.

Hill, J. (2001). *Building Community in Web-Based Learning Environments: Strategies and Techniques.* Retrieved on 2 May 2002 from the World Wide Web: www.ausweb.scu.edu.au/aw01/papers/refereed/hill/paper.html.

Hiltz, S.R. (1996). The virtual classroom: using computer mediated communications for university teaching. *Journal of Communication, 36* (2), 51-53.

Jackson, R.H. (2001). *Defining e-Learning - Different Shades of "Online".* Web Based Learning Resources Library. [Online]. Retrieved on 2 May 2002 from the World Wide Web: http://www.outreach.utk.edu/weblearning/.

Jones, V., Jo, J. & Crantich, G. (2000). A study of students' responses to WBI within a traditional learning environment. The sixth Australian World Wide Web Conference. [Online]. Retrieved on 17 April 2002 from the World Wide Web:: http://ausweb.scu.edu.au/aw2k/papers/jones/paper.html. Accessed: 17 April 2002.

Laurillard, D. (1993). *Rethinking University Teaching: A Framework for the Effective Use of Educational Technology.* London: Routledge.

Marold, K. (2002). The 21$^{st}$ century learning model: electronic tutelage realized. *Journal of Information Technology Education, 1* (2), 113 – 123.

Nasseh, B. (1996). A Study of Computer-Based Distance Education in Higher

Copyright © 2003, Idea Group Inc. Copying or distributing in print or electronic forms without written permission of Idea Group Inc. is prohibited.

Education Institutions in Indiana. Retrieved on 24 April 2002 from the World Wide Web: www.bsu.edu/classes/nasseh/study/research.html.

Ndahi, H. (1999). Utilization of distance learning technology among industrial and technical teacher education faculty. *Journal of Industrial Teacher Education, 36* (4).

O' Keefe, S. & McGrath, D. (2000). On-line delivery in higher education: What questions should we be asking? ACILITE 2000. [Online]. Retrieved on 24 March 2002 from the World Wide Web:http://www.ascilite.org.au/conferences/coffs00/papers/suzanne_okeefe.pdf.

Passerini, K. & Granger, M. (1999). A developmental model for distant learning using the Internet. *Computers & Education, 34,* (1-15).

Pennel, R. (1996). Managing online learning. The Second Australian World Wide Web Conference. [Online]. Retrieved on 12 March 2002 from the World Wide Web: http://ausweb.scu.edu.au/aw96/educn/pennell/paper.htm.

Piccoli, G., Ahmad, R. & Ives, B. (2001). Web-based virtual learning environments: A research framework and a preliminary assessment of effectiveness in basic IT skills training. *MIS Quarterly, 25* (4), 401-426.

Retalis, S. & Avgeriou, P. (2002). Modeling web-based instructional services. *Journal of Information Technology Education, 1* (1), 25–41.

Robson, J. (2000). Evaluating on-line teaching. *Open Learning, 15* (2), 153-157.

Roderic, A. (2002). Online learning – Introduction. *VNU Business Media.* [Online]. Retrieved on 17 July 2002 from the World Wide Web: http://www.vnulearning.com/wp/introduction.htm.

Rogers, P. (2001). Traditions to Transformations: The Forced Evolution of Higher Education. *Educational Technology Review.* [Online]. Retrieved from the World Wide Web: www.aace.org/pubs/etr/rogers.cfm.

Rossiter, D. (1999). Building a web-based framework to embed the teaching and learning of technological literacies. *The Fifth Australian World Wide Web Conference,* Southern Cross University. [Online]. Retrieved on 18 April 2002 from the World Wide Web:.http://ausweb.scu.edu.au/aw99/papers/rossiter/paper.html.

Schulze, S. (2000). The online learning environment of students: Educators' Perspectives. *Journal of Education and Training, 21* (1), 25 – 38.

Serdiukov, P. (2001). Models of distance higher education: Fully automated or partially human? *Educational Technology Review.* [Online]. Retrieved on 19 February 2002 from the World Wide Web:.ww.aace.org/pubs/etr/serdiukov.cfm.

Sherson, G. (1996). Teaching using the World Wide Web. *Universal College of Learning.* [Online]. Retrieved on 19 February 2002 from the World Wide

Copyright © 2003, Idea Group Inc. Copying or distributing in print or electronic forms without written permission of Idea Group Inc. is prohibited.

Web: http://www.ucol.ac.nz/~g.sherson/papers/webdelivery.htm.

Sherson, G. (1999). Education at the digital campus. *Universal College of Learning*. Retrieved from the World Wide Web: www.ucol.ac.nz/~g.sherson/papers/digital.htm.

Simpson, O. (2000). *Supporting Students In Open and Distance Learning*. London: Kogan Page.

Slay, J. (1997). The use of the Internet in creating an effective learning environment. *The Third Australian World Wide Web Conference*, Southern Cross University. [Online]. Retrieved on 17 April 2002 from the World Wide Web: http://ausweb.scu.edu.au/aw97/papers/slay/paper.htm.

Taylor, R. (2002). Pros and cons of online learning – A faculty perspective. *Journal of European Industrial Training , 26* (1), 24 – 37.

Tian, S. (2001). The World Wide Web: A vehicle to develop interactive learning and teaching applications. *Internet Research: Electronic Networking Applications and Policy, 11* (1) 74–83.

Tiffin, J. & Rajasingham, L. (1995). In search of the virtual class. London: Routledge.

Uys, P.M. (1999). Towards the virtual class: Technology issues from fractal management perspective. *"ED-MEDIA 99"* Conference in Seattle, USA.

Webster. J. & Hackley, P. (1997). Teaching effectiveness in technology-mediated distance learning. *Academy of Management Review, 40* (6), 1282–1309.

Wells, J. (2000). Effects of an on-line computer-mediated communication course, prior computer experience and internet knowledge, and learning styles on students' internet attitudes computer-mediated technologies and new educational challenges. *Journal of Industrial Teacher Education*. [Online]. Retrieved on 19 October 2002 from the World Wide Web: http://scholar.lib.vt.edu/ejournals/JITE/v37n3/wells.html.

Zepke, N. (1998). Instructional design for distance delivery using hypertext and the internet: assumptions and applications. *Quality in Higher Education, 4* (2), 173-186.

# GLOSSARY

**Asynchronous** – Involves a sequence of operations without a regular predicate time relationship. Thus operations do not happen at regular timed intervals, but an operation will begin only after a previous operation is completed.

**Computer-based courses** – The process of teaching and learning is designed based on the capabilities of computer and telecommunication, including course content.

Copyright © 2003, Idea Group Inc. Copying or distributing in print or electronic forms without written permission of Idea Group Inc. is prohibited.

**Communication Tools** – It can be anything from Internet to electronic mail that is used as a method of communication between lecturer and student.

**Delivery Tools** – Used as methods of delivering notes, exams or assignments (e.g., File Transfer Protocol and bulletin boards).

**Distance Education** – The process that delivers interactive and responsive learning opportunities to learners at a time, place, and in the form appropriate and convenient to learners (Roger Kaufman, 1995).

**Electronic Learning (E-learning)** – using the Internet environment to contribute to the learning process.

**File Transfer Protocol** – In conjunction with the proper local software, FTP allows computers connected to the Internet to exchange files, regardless of the type of computer software.

**Internet Relay Chat** – Synchronous communication tool that allows people all over the Internet to communicate with one another in real-time

**Network** – Interconnected system of computers, terminals and communications channels and devices.

**Synchronous** – Synchronous data communication requires that each end of an exchange of communication respond in turn without initiating a new communication. A typical activity that might use a synchronous protocol would be a transmission of files from one point to another. As each transmission is received, a response is returned indicating success or the need to resend. Each successive transmission of data requires a response to the previous transmission before a new one can be initiated.

**Telecommunications** – The sending of information in any form (e.g., voice, data, text, images etc.) from one place to another, using electronic media.

**Virtual Class** – The learning process enabled solely by telecommunications.

**World Wide Web (WWW)** – It is an information system on the Internet based on hypertext that offers a great power for an information search and delivery system. The current feature of WWW includes text, sound, graphics and video.

Copyright © 2003, Idea Group Inc. Copying or distributing in print or electronic forms without written permission of Idea Group Inc. is prohibited.

# APPENDIX

**Information Systems: NB_Survey_August2002**

### Question: Q1
Gender:
1.    Male        2.  Female

### Question: Q2
Age group:
1.    Under 21    2.   21 - 25      3.   26 - 30      4.   Older than 30

### Question: Q3
Which course are you currently registered for?
1.    Information Technology        2.    Financial Information Systems

### Question: Q5
Did you have any computer experience before doing the e-learning subject?
1.    Yes      2.  No

### Question: Q6
How often do you use the Internet?
1.    1 - 5 times a month        2.   6 - 10 times a month
3.    11 - 20 times a month      4.   more than 20 times a month

### Question: Q7
In general, you enjoy using a computer
    1.   Strongly disagree        2.   Disagree        3.   Neutral
    4.   Agree                    5.   Strongly Agree

### Question: Q8
You prefer traditional (face-to-face) lectures to online lectures
    1.   Strongly disagree        2.   Disagree        3.   Neutral
    4.   Agree                    5.   Strongly Agree

### Question: Q9
Online lectures allow you, more conveniently, to look up the information you need.
1.    Strongly disagree        2.   Disagree        3.   Neutral
4.    Agree                    5.   Strongly Agree

Copyright © 2003, Idea Group Inc. Copying or distributing in print or electronic forms without written permission of Idea Group Inc. is prohibited.

## Question: Q10
Memorising certain types of information is easier in the online environment.

1.   Strongly disagree      2.   Disagree      3.   Neutral
4.   Agree                  5.   Strongly Agree

## Question: Q11
Learning is more fun with online lectures.

1.   Strongly disagree      2.   Disagree      3.   Neutral
4.   Agree                  5.   Strongly Agree

## Question: Q12
You can easily reference any section of the lecture in an online environment.

1.   Strongly disagree      2.   Disagree      3.   Neutral
4.   Agree                  5.   Strongly Agree

## Question: Q13
Communication between student and facilitator is crucial in the e-learning process.

1.   Strongly disagree      2.   Disagree      3.   Neutral
4.   Agree                  5.   Strongly Agree

## Question: Q14
Interaction among students is crucial to the e-learning process.

1.   Strongly disagree      2.   Disagree      3.   Neutral
4.   Agree                  5.   Strongly Agree

## Question: Q15
The fact that I frequently have to find and/or use a computer to participate in this course is a source of annoyance to me

1.   Strongly disagree      2.   Disagree      3.   Neutral
4.   Agree                  5.   Strongly Agree

## Question: Q16
I received sufficient instruction on the use of the e-learning programs (e.g., WebCT) for this course.

1.   Strongly disagree      2.   Disagree      3.   Neutral
4.   Agree                  5.   Strongly Agree

Copyright © 2003, Idea Group Inc. Copying or distributing in print or electronic forms without written permission of Idea Group Inc. is prohibited.

**Question: Q17**
The amount of time required for computer use in this course is excessive.
1.    Strongly disagree        2.  Disagree        3.  Neutral
4.    Agree                    5.  Strongly Agree

**Question: Q18**
My computer literacy was adequate for performing the functions required of the e-learning tools. (e.g., chat, bulletin boards)
1.    Strongly disagree        2.  Disagree        3.  Neutral
4.    Agree                    5.  Strongly Agree

**Question: Q19**
Using e-learning as a platform for this subject has increased my course workload.
1.    Strongly disagree        2.  Disagree        3.  Neutral
4.    Agree                    5.  Strongly Agree

**Question: Q20**
E-learning helped me to learn my subject content more quickly.
1.    Strongly disagree        2.  Disagree        3.  Neutral
4.    Agree                    5.  Strongly Agree

**Question: Q21**
E-Learning has helped me develop new computing skills.
1.    Strongly disagree        2.  Disagree        3.  Neutral
4.    Agree                    5.  Strongly Agree

**Question: Q22**
I prefer having the subject material online than using a textbook.
1.    Strongly disagree        2.  Disagree        3.  Neutral
4.    Agree                    5.  Strongly Agree

**Question: Q23**
The Bulletin Board is effectively used in this subject
1.    Strongly disagree        2.  Disagree        3.  Neutral
4.    Agree                    5.  Strongly Agree

Copyright © 2003, Idea Group Inc. Copying or distributing in print or electronic forms without written permission of Idea Group Inc. is prohibited.

**Question: Q24**
The Private Mail is effectively used in this subject
1.   Strongly disagree        2.   Disagree        3.   Neutral
4.   Agree                        5.   Strongly Agree

**Question: Q25**
The Chat Room is effectively used in this subject
1.   Strongly disagree        2.   Disagree        3.   Neutral
4.   Agree                        5.   Strongly Agree

**Question: Q26**
The Calendar is effectively used in this subject
1.   Strongly disagree        2.   Disagree        3.   Neutral
4.   Agree                        5.   Strongly Agree

**Question: Q27**
The Downloads tool is effectively used in this subject
1.   Strongly disagree        2.   Disagree        3.   Neutral
4.   Agree                        5.   Strongly Agree

**Question: Q28**
The Discussions tool is effectively used in this subject
1.   Strongly disagree        2.   Disagree        3.   Neutral
4.   Agree                        5.   Strongly Agree

**Question: Q29**
The Self-Test tool is effectively used in this subject
1.   Strongly disagree        2.   Disagree        3.   Neutral
4.   Agree                        5.   Strongly Agree

**Question: Q30**
The Quiz tool is effectively used in this subject
1.   Strongly disagree        2.   Disagree        3.   Neutral
4.   Agree                        5.   Strongly Agree

**Question: Q31**
The Home Page tool is effectively used in this subject
1.   Strongly disagree        2.   Disagree        3.   Neutral
4.   Agree                        5.   Strongly Agree

Copyright © 2003, Idea Group Inc. Copying or distributing in print or electronic forms without written permission of Idea Group Inc. is prohibited.

**Question: Q32**

The My Progress tool is effectively used in this subject

| | | | |
|---|---|---|---|
| 1. | Strongly disagree | 2. Disagree | 3. Neutral |
| 4. | Agree | 5. Strongly Agree | |

**Question: Q33**

The My Grades tool is effectively used in this subject

| | | | |
|---|---|---|---|
| 1. | Strongly disagree | 2. Disagree | 3. Neutral |
| 4. | Agree | 5. Strongly Agree | |

**Question: Q34**

The Assignments tool is effectively used in this subject

| | | | |
|---|---|---|---|
| 1. | Strongly disagree | 2. Disagree | 3. Neutral |
| 4. | Agree | 5. Strongly Agree | |

**Question: Q35**

The Search tool is effectively used in this subject

| | | | |
|---|---|---|---|
| 1. | Strongly disagree | 2. Disagree | 3. Neutral |
| 4. | Agree | 5. Strongly Agree | |

Copyright © 2003, Idea Group Inc. Copying or distributing in print or electronic forms without written permission of Idea Group Inc. is prohibited.

## Chapter XI

# Relating Cognitive Problem-Solving Style to User Resistance

Michael Mullany
Northland Polytechnic, New Zealand

Peter Lay
Northland Polytechnic, New Zealand

## ABSTRACT

*This study investigated the relationships between user resistance to new information systems and the differences in cognitive problem-solving styles between systems developers (analysts) and users. In addition, associations were tested between user resistance and the following: system accuracy, system reliability, the analyst's attitude, the analyst-user relationship, analyst-user dissonance, the user's age and the user's length of service with his current employer.*

*A significant positive association between user resistance and analyst-user cognitive style difference was found. A model was then developed which enables the estimation of user resistance prior to system development with the aid of the Kirton Adaptive/Innovative Inventory (an instrument which measures cognitive style). Significant negative associations were found to exist between user resistance and system accuracy, and user resistance and system reliability. No relationships between user resistance and either user age or user length of service were found.*

Copyright © 2003, Idea Group Inc. Copying or distributing in print or electronic forms without written permission of Idea Group Inc. is prohibited.

# INTRODUCTION

There are conflicting views regarding the influence of cognitive style on user resistance. Hirschheim and Newman (1988)[1], for example, note that a user may resist a system because the mode of presentation of that system does not match the user's cognitive style. In other words, the system does not follow a problem-solving sequence that is totally acceptable to the user. Huber (1983), however, deduced from his own literary survey that cognitive style theory had not, and probably would never, provide guidelines for system design. Nonetheless, a careful analysis of Huber's study reveals that it merely rejects attempts to match a system to a given user's cognitive style. It does not reject attempts to match the systems developer cognitively with a given user. On the contrary, Huber's study encourages the use of cognitive style theory for personnel selection and placement purposes. Since little direct evidence was found in the IS literature of previous attempts to match systems developers and users cognitively, it was decided to investigate user resistance along this more novel line.

The purpose of this study, then, was to test relationships between user resistance to computer-based systems, including those commonly associated with e-Commerce and cognitive style. This term, or the alternative, cognitive problem-solving style is used to denote an individual's approach to problem solving. The Kirton Adaption-innovation Inventory (KAI) was selected as an instrument to measure cognitive styles. The KAI, based on Adaption-innovation (A-I) theory, measures an individual's preference for either an "adaptive" or "innovative" approach to problem solving. Adaptive problem-solvers tend to follow prescribed and traditional methods, whilst innovators seek new and often unexpected solutions. These concepts will be expanded on later.

# SUMMARY OF PRIOR RESEARCH

The study covered four areas of prior research. These are: IS user resistance to new systems; resistance to change in other areas not necessarily related to IS (e.g., e-commerce); the influence of cognitive problem-solving styles on resistance to change; and measures of system success and their application. Since e-commerce success is largely a matter of user opinion, the aspect of user satisfaction and its measurement is covered under the discussion of the last area. A link is claimed between user satisfaction (or dissatisfaction) and user resistance, thus the possibility of using user satisfaction as a converse substitute measure of user resistance is considered.

Copyright © 2003, Idea Group Inc. Copying or distributing in print or electronic forms without written permission of Idea Group Inc. is prohibited.

## User Resistance - IS Studies and E-Commerce

Only five studies of significance were found in the IS literature: those of Keen (1981), Hirschheim and Newman (1988), Bruwer (1984), Markus (1983) and De Brabander and Thiers (1984). The studies of both Keen and Hirschheim and Newman are literature surveys. The former concludes that new systems that represent radical change, as opposed to those that cause incremental or evolutionary change, will be avoided or resisted. Also, since the redistribution of data caused by a new information system is a political resource, the interests of certain groups will be affected. In the latter, the authors define resistance as an adverse reaction to a proposed change, which may be overt or covert. They suggest that the impact of user resistance emerging during implementation may take any of the following forms: low productivity, low effectiveness, high labour turn- over, disputes, absenteeism, psychological withdrawal, aggression, sabotage of machinery, the system being blamed for all difficulties experienced (including incorrect data entries), and lack of management support for the system.

Bruwer (1984) studied resistance to computerization in a single organization, where 140 computerized systems, used by about 1,200 clerical staff and 114 managers, were investigated. He claims that older managers with longer experiences are more negative towards computerization than are younger ones.

Markus (1983) identified three general theories explaining user resistance from the IS literature, which she then assessed in the light of a single case study. The following were identified as causing resistance: internal individual or group factors, such as a non-analytic cognitive style; factors inherent in the system under implementation, such as technically deficient systems or systems which are not user-friendly; and certain interactions between factors inherent in the user and others intrinsic to the system. In the main, user resistance behaviors took the form of complaints against the system that were considered unfair. This, it will be noted, corroborated the behaviors identified by Hirschheim and Newman.

Finally, De Brabander and Thiers (1984) studied certain defective implementation behaviors in the form of not adhering to plans, which resulted in reduced efficiency of task-accomplishment amongst users. They concluded that the reason that their users did not adhere to plans was that they were subject to the sanctioning powers of the corresponding IS specialists. This effect, they noted, was enhanced by the presence of a semantic gap that they define as the employment of differing conceptual definitions for aspects of the same task.

As can be seen from the above, there has been little of a definitive nature in the IS literature pertaining to user resistance. None of the work described above attempted to associate user resistance with any aspects of cognitive style. This

Copyright © 2003, Idea Group Inc. Copying or distributing in print or electronic forms without written permission of Idea Group Inc. is prohibited.

prompted an examination of studies that had been conducted on resistance in related fields.

## Resistance to Change in other Areas

Rosen and Jerdee (1976) conducted an enquiry into age stereotypes of employees, which suggests that increased resistance to change is a generally *believed* characteristic of older employees. They found that there was a tendency for younger persons to be considered more receptive to new ideas, and older persons, by implication, more resistant to them. After a further intensive literature survey, however, they found that there was little research evidence to support such beliefs, and concluded that older employees are the potential victims of unjustified discrimination. This study, it should be noted, contradicts the findings of Bruwer.

In a study of the impact of manpower-flows on innovation, from which inferences for resistance to change can be drawn, Ettlie (1985) offers support for the premises that innovation is aided by: an influx of new employees (if not too disruptive); the degree to which the organizational structure is decentralized; the complexity of the organization; and the availability of slack resources. Hence, by implication, the same factors must ameliorate resistance to innovation.

## Cognitive Problem-Solving Styles and Resistance to Change

The third area of analysis is that of cognitive problem-solving style and its impact on system success (e.g., e-commerce). If the way in which analysts define and describe tasks can be implicated in the phenomenon of user resistance, then it stands to reason that the cognitive approaches, or problem-solving styles of the analyst and user are also significant. This follows from the fact that definition and description are basic to cognitive information processing. Hence the implication of the semantic gap also suggests the role of cognitive differences between the user and the analyst in user resistance (Zmud, 1983).

There have been numerous studies attempting to categorize problem-solving style, such as those of Woodruff (1980) and Lusk and Kersnick (1970), but one of the major research efforts in this area has been performed by Kirton (1976, 1980, 1984, 1985, 1987, 1988) under what he terms Adaption-Innovation theory (Kirton, 1976). Adaption-Innovation theory, Kirton claims, explains in an empirical way many of the anomalies surrounding resistance to change in occupational situations. He proposes that any person can be located on a continuum ranging between two extreme cognitive styles; from an ability to adapt existing technologies (an adaptor) to an ability to use new or different technologies (an innovator). This

Copyright © 2003, Idea Group Inc. Copying or distributing in print or electronic forms without written permission of Idea Group Inc. is prohibited.

proposition, he claims, is relevant to the analysis of organizational change, and offers new insights into the concept of resistance to change. Kirton's enquiry into the blocking of new initiatives in several large corporate institutions offers a fundamental conclusion regarding resistance to change. In general, a person will exhibit less resistance to ideas put forward by another of similar cognitive style. He further discusses the notion of an "agent for change", and stresses that it is not necessarily associated with innovation. An agent for change is rather a person "who can successfully accept, and be accepted into, an environment alien to his own", or ''as a competent individual who has enough skill to be successful in a particular environment" (Kirton, 1984).

The Kirton Adaption-innovation Inventory (KAI) instrument provides a means for measuring individuals' cognitive problem-solving styles (Kirton, 1987). These scores are stable, and little variation is reported with time or age. This instrument rates innovators with higher scores than adaptors. These scores have relative as well as absolute significance; hence if A's KAI score is higher than E's, it is meaningful to describe A as "more innovative" than E, and B as "more adaptive" than A.

As a final point, Huber (1983) in a literary survey concludes that the literature available on cognitive style research in IS does not support a satisfactory basis for recommendations on IS design. Also, he expresses pessimism that this approach will ever provide such guidelines. However, he does recommend cognitive style research in three areas: career counseling, personnel selection and placement, and coaching and training. The second of these has implications for the placing of the most suitable analyst with a given user, or vice versa. This, together with the previous discussion of A-I theory, suggests that the KAI might be used for matching users and analysts successfully. It was thus obvious to select the KAI as the instrument to measure cognitive style in this study.

## Measures of System Success and User Satisfaction

The final research area to be analyzed was that of system success and the related area of user satisfaction. Attempts are found in the literature to identify those attributes of systems that tend to satisfy (and/or dissatisfy) users. In one of the most comprehensive analyses of user satisfaction instruments, Ives, Olson and Baroudi (1983) conducted a psychometric analysis of four "User Information Satisfaction" instruments. These were, Gallagher's questionnaire, Jenkins and Rickett's 20-item measure, Larcker and Lessing's perceived usefulness instrument, and Pearson's 39-factor instrument. In conclusion, they favoured Pearson's instrument as the most predictive and exhibiting the greatest construct validity. The shortened version of Pearson's instrument that they subsequently developed, they considered a prom-

Copyright © 2003, Idea Group Inc. Copying or distributing in print or electronic forms without written permission of Idea Group Inc. is prohibited.

ising advance in the measure of user satisfaction. In their own research, the five of the thirty-nine factors found to be most significant by Bailey and Pearson (1983) were: accuracy (correctness of output), reliability (dependability of output), timeliness (output available at a time suitable for use), relevancy of output, and confidence in the system.

There have been various other studies of user involvement or user satisfaction (Eveland, 1977; Olson & Ives, 1981; Robey & Farrow, 1982; Rushinek & Rushinek, 1986), but the conclusions reached do not corroborate or complement each other.

# DEVELOPMENT AND STATEMENT OF HYPOTHESES

The studies by Lusk and Kersnick, Kirton, and Hirschheim and Newrnan suggest the important role of cognitive styles in occupational situations. As was noted, in general, a person will exhibit less resistance to ideas put forward by another person of similar cognitive style. This in turn justifies the submission that user resistance is associated with differences in developer-user cognitive problem-solving styles. This was the basis of the primary hypothesis of this study. The research also attempted to build on the work of other IS researchers by using the instruments of Bailey and Pearson (1983) and Rushinek and Rushinek (1986), and challenging the conclusions of Bruwer (1984).

In the light of the foregoing discussion, this study poses the following central questions:

i.    During system development, implementation and maintenance, is there a relationship between user resistance and cognitive styles (or cognitive style differences) associated with the analyst-user interface?

ii.    Can cognitive style theory be used to predict certain general aspects and behaviors of a given analyst-user interface during the development and implementation of an information system?

iii.    Are there factors related to systems or their manner of implementation that are associated with user resistance?

iv.    Do the ages and lengths of service of analysts and/or users play a role in user resistance?

Consequences of the literature-based discussions above lead directly and indirectly to certain hypotheses regarding the causes of user resistance, which fall into the above categories. Before such hypotheses could be developed, however, a reliable instrument to measure user resistance needed to be selected.

Copyright © 2003, Idea Group Inc. Copying or distributing in print or electronic forms without written permission of Idea Group Inc. is prohibited.

## The Choice of an Instrument to Measure User Resistance

Had the link between user satisfaction and user resistance been assumed, an instrument which measures user satisfaction could have been chosen to yield a converse substitute measure for user resistance. Pearson's instrument for measuring user satisfaction has been acclaimed as one of the best in the IS literature to date. However, its length (and even the length of the modified version suggested by Ives, et al.) must raise criticism. If the respondent were to complete and return the form on a voluntary basis, then the sample gathered would tend to contain the more conscientious respondents; namely, those who are prepared to fill in and return lengthy questionnaire forms. Since conscientiousness is claimed by Kirton (1984) to be associated with an adaptive cognitive style, the risk of bias is immediately evident with this questionnaire. An instrument other than a self-report questionnaire of the above type was thus indicated.

Ives, et al (1983) developed the notion of substitute or "surrogate" measures for entities not capable of direct measurement themselves. In the light of this and the reservations just expressed, an alternative instrument was devised, which constituted a quantified expression of dissatisfaction to a person independent of the user's organization. The potential weakness inherent in allowing respondents to complete and return their own forms was removed by collecting data at personal, confidential interviews with the users. The user of each system under investigation was asked to enumerate all the problems that he considered or heard had occurred during the implementation or early life of the system. He was effectively invited, in confidence, to make complaints against the system and/ or its manner of development and implementation. Each complaint was recorded, and then the user was asked to weight his complaints in terms of severity on a seven-point scale (see Appendix). The sum of the weights for each complaint was taken as the user's "resistance score" or "R-score".

The validity of the R-score is argued on the basis that user resistance can only be exhibited in one of four ways: by what the user says or by what the user does (overt resistance), or by what he doesn't say or doesn't do (covert resistance). Anyone (or any combination) of these four phenomena, if measured and quantified, would be expected to provide an observable measure of user resistance. Furthermore, as an independent party is not part of the political structure of the organization, comments made to him about the system are likely to be more objective than hearsay filtering back to the management.

However, there are two issues that require clarification. Firstly, there is the question of whether or not overt user dissatisfaction, expressed in the form of complaints, is a legitimate measure of user resistance. Both the studies by Hirschheim and Newman (1988) and Markus (1983) suggest a significant associa-

Copyright © 2003, Idea Group Inc. Copying or distributing in print or electronic forms without written permission of Idea Group Inc. is prohibited.

tion between user resistance and user complaint. The former study asserts that amongst the various types of user resistance, there are: aggression; and the system being blamed for all problems prevailing during implementation, including incorrect data entries. This is confirmed in the latter study, where user resistance was observed mainly to take the form of complaints, many of which Markus considered unfair. These literary sources thus support a method that quantifies user dissatisfaction as a substitute measure for user resistance.

However, there are types of resistance that cannot easily be proved to be associated with complaint. Examples of such are psychological withdrawal and absenteeism, because these are types of covert resistance whilst a complaint is a case of overt resistance. The answer to the first question, then, is that the study must be limited to investigating only those forms of resistance that are measurable in terms of overt expressions of dissatisfaction. Whilst this may at first seem restrictive, it should be noted that perceived success of the system to management is most likely to be tied to reports of success from its employee users. In other words, this measurement technique will appeal to managers whose interests are largely the warding off of complaints against new systems from their staff.

The second issue is the question of how complete the list of the user's complaints would be, or alternatively, what the significance would be if the user forgot to itemize certain problems. It is argued that this is not a significant issue, since the R-score method aims to observe the user in the *process of complaining*. If, as noted above, users tend to make unfair complaint as a resistance behavior, then users will tend to invent or exaggerate complaints as an expression of their level of dissatisfaction. Quite clearly, the issue of whether or not the user remembers all the real problems that occurred during implementation then becomes less relevant.

The R-score approach does, however, prescribe certain stipulations that were borne in mind during the research design. Firstly, the users needed to be interviewed personally by a researcher to obtain their views of systems. This follows immediately from the need to relieve the respondent of the obligation to complete and return a form, as discussed above. Secondly, every effort had to be made to convince the user that all his responses were to be kept confidential; particularly from the analyst(s), management and potential business rivals. Without this stipulation, the advantage of having opinions expressed to an independent researcher (as mentioned above) would have been lost.

The weighting of the severity of the complaints is based on a model developed by Wanous and Lawler (1972) for measuring a worker's job satisfaction. According to this model, an employee's overall job satisfaction is the weighted sum of his satisfaction with all significant facets of the job. This can be expressed algebraically as:

Copyright © 2003, Idea Group Inc. Copying or distributing in print or electronic forms without written permission of Idea Group Inc. is prohibited.

$$S = \sum_{i=1}^{n} s_i w_i,$$

where:

S = respondent's overall satisfaction with n significant facets of the job,
$s_i$ = respondent's satisfaction with facet i, rated on a seven-point scale, and
$w_i$ = the importance of facet i to the respondent, rated on a seven-point scale.

Pearson made use of this model in his 39-factor instrument discussed earlier (Bailey & Pearson, 1983). However, the R-score method makes three fundamental departures in its application of the same. Firstly, it assumes that weighted responses to dissatisfaction with various facets of a system and its manner of implementation will give a valid measure of overall dissatisfaction. It is difficult to see why all the arguments which apply to weighted measures of satisfaction cannot also apply to its antipode, dissatisfaction, hence this variation was assumed valid. Secondly, the R-score method assumes that overall dissatisfaction gives a valid measure of its surrogate, user resistance. This stems directly from the discussion above so long as it is understood that only those forms of resistance variation were assumed valid. Secondly, the R-score method assumes that overall dissatisfaction gives a valid measure of its surrogate, user resistance. This stems directly from the discussion above so long as it is understood that only those forms of resistance measured by overt expressions of dissatisfaction are intended. Finally, the facets of dissatisfaction are not pre-specified, but are enumerated by the respondent (that is, the user) himself. This was justified in terms of Wanous and Lawler's model by rating any complaint raised by the user with unit importance (and, of course, those not raised by the user with zero importance). This meant that the R-score could be measured simply by summing the user's dissatisfaction ratings for his own complaints against the system. In mathematical terms, the R-score, R, after the user has made n complaints in respect of the system and/or its manner of implementation, can be expressed as:

$$R = \sum_{i=1}^{n} s_i,$$

where $s_i$ is defined as the severity of complaint i to the user, rated on the following seven-point scale:
(7)   a totally unsolvable problem
(6)   a very serious problem

Copyright © 2003, Idea Group Inc. Copying or distributing in print or electronic forms without written permission of Idea Group Inc. is prohibited.

(5)   a serious problem
(4)   a rather serious problem
(3)   a significant problem
(2)   a slight problem
(1)   not really a problem

In the light of the previous discussion, the R-score was assumed to be a valid measure of user resistance, despite its novelty. Of course it can be argued that the R-score might change with the nature of the researcher conducting the interview. It was beyond the scope of this study to investigate such a conjecture, but since it was sufficient to show resistance in the relative sense only (comparing resistance between systems investigated by the same person), this criticism was not considered significant.

## The Role of Cognitive Problem-Solving Styles

A submission based on A-I theory, made earlier was that user resistance is associated with differences in developer-user cognitive problem-solving styles. This can now be stated as the following hypothesis:

*HI: The user's R-score is positively associated with the absolute analyst-user KAI score difference for a given information system*

Further hypotheses were thus formulated for testing. For example, a user should tend to see an analyst who is more innovative than himself spend surprisingly little time studying the system requirements. This follows from the tendency of innovators to not to wed themselves too long to any system, and to seek continued novelty of activity. Since the user will consider the analyst an "expert" (lacking in the beginning a frame of reference to consider him anything else), he should also assume that the analyst has absorbed all the details in this surprisingly short time. The degree to which a user will see the analyst as more innovative or adaptive than himself is measurable as the algebraic difference between their KAI scores. Hence the hypothesis:

*H2(a): The analyst-user KAI score difference is positively associated with the user's seven- point rating of how quickly the analyst absorbed (grasped) the system requirements*

According to Kirton (1984), innovators tend to pursue a course of action with more apparent certainty than do adaptors. This leads to the obvious conjecture that

Copyright © 2003, Idea Group Inc. Copying or distributing in print or electronic forms without written permission of Idea Group Inc. is prohibited.

an innovative analyst, when dealing with an adaptive user, will tend to follow his own ideas rather than to pay over-much attention to the user. The adaptor, prone to high self-doubt, would be expected to give in to the analyst's views. Hence:

*H2(b): The analyst-user KAI score difference is positively associated with the user's seven-point rating of the extent to which the analyst followed his own ideas and/or ignored the user's opinions*

Innovators, with the predilection for novelty, would be expected to add to the system development effort extraneous features, which the more adaptive user would consider unnecessary and time wasting. This suggested the hypothesis:

*H2(c): The analyst-user KAI score difference is positively associated with the user's seven-pointrating of the extent to which the analyst wasted time on peripheral issues*

Innovators are often seen as abrasive, creating dissonance. Hence, a measure of dissonance would be expected to exist between individuals who differ markedly in cognitive style. In addition, Bailey and Pearson (1983) found that a user's relationship with the IS staff, as well as the attitude of the IS staff, influenced user satisfaction. By "attitude", Bailey and Pearson were referring to the analyst's willingness to assist the user. It is doubtful that a user would report friction between himself and the analyst, and then describe their relationship as sound, and the analyst as helpful. Consequently, the user's ratings of these three factors should be highly correlated. To test this, the following hypothesis was specified:

*H2(d): The absolute analyst-user KAI score difference is negatively associated with the user's seven-point rating of his relationship with the analyst*

The key factors thought by Pearson to play a role in user satisfaction were identified earlier. As it is of interest to see how their approaches compared with that of this study, a short-list of user-satisfaction factors was drawn up. Negative associations were then postulated between these factors and the R-score. The hypotheses tested in respect of the above were thus:

*H3(a): There is a negative association between the user's R-score and his seven-point rating of the system's level of accuracy*

Copyright © 2003, Idea Group Inc. Copying or distributing in print or electronic forms without written permission of Idea Group Inc. is prohibited.

*H3(b): There is a negative association between the user's R-score and his seven-point rating of the degree he can rely on the system's output information*

## The Roles of Age and Length of Service

After examination of the study by Bruwer, beliefs were identified that older and/or more experienced users in the organization will be most resistant to new computer systems. These beliefs infer the following hypotheses:

*H4(a): A user's R-score for a given information system is positively associated with his age*

*H4(b): A user's R-score for a given information system is positively associated with his length of service in his organization*

*H4(c): A user's R-score for a given information system is positively associated with the absolute difference between his age and the analyst's age*

*H4(d): A user's R-score for a given information system is positively associated with the absolute difference between his and the analyst's length of service in the same organization*

# THE RESEARCH METHODOLOGY AND DESIGN

Ten organizations in the Cape Town and Johannesburg areas in South Africa participated. Within these organizations, 34 live (post-implementation) systems were randomly selected. These systems had all been in operation for between two and 26 months, and in each case, both a key analyst and key user could be identified.

First the analyst was interviewed, since he could supply details of the user; most particularly, where the key user could be located. This information was recorded together with the analyst's age and length of service. At the end of this interview, the KAI was administered to the analyst. Every effort was made to ensure that standard testing conditions prevailed during the administration of the KAI. Following this, the user was interviewed, whereupon a System Satisfaction Schedule (see Appendix) was completed. At the end of the interview, the KAI was administered to the user. The procedure and circumstances of the user interviews and KAI administration were similar to those described above for the analysts.

Copyright © 2003, Idea Group Inc. Copying or distributing in print or electronic forms without written permission of Idea Group Inc. is prohibited.

Of significance is the question of whether or not complaints were distinguished in a consistent manner. With this in mind, notes were carefully taken at each interview of any unusual occurrences and responses, and how they were handled. Similar action was thus ensured where similar circumstances arose subsequently. For example, experience soon showed that responses to seven-point scales required special handling, because several of the respondents could not remember all the options at once. Even though the respondent was shown the form, he often required help in making his assessment. This assistance was given in the form of a two-tiered approach. First, the respondent was asked to make a crude assessment out of options 2, 4 and 6, and then he was asked to refine his choice by making a selection within one of his original choice. For example, if he first chose option 6, he would then be asked to make a final assessment out of 5, 6 or 7.

The criticism that points of complaint were recorded to correlate with KAI scores and/or differences was counteracted in the research design as follows: the user R-score was measured close to the beginning of the interview, before the user's KAI score was measured, and before it became possible to make a reliable guess of the user's KAI score. Other questions on the System Satisfaction Schedule were not asked until after the R-score had been measured in the case of a user, once again to avoid guesses as to the user's cognitive style.

# FINDINGS AND DISCUSSION

In this study, thirty-four systems were researched, hence the data is divided into several univariate samples of thirty-four (that is, sample-size $n = 34$ in all cases).

The levels of significance employed in this study were based on the opinions found in human science and statistical literature. These opinions are summarized in Table 1. Based on the opinions of respected experts as recorded in Table 1, the qualitative ratings listed in Table 2 were assumed for this study. The data was then stratified into those univariate samples identified in Table 3.

The mean and standard deviation was calculated for each sample, which was then tested for goodness-of-fit to the normal distribution.

## The Randomness of the Data Samples

The means for the analyst and user KAI score samples were respectively, 102.9 and 101.6, while their respective standard deviations are 12.55 and 14.09. Kirton's British sample of KAI scores for 562 persons exhibited a normal distribution with mean 95 and standard deviation 18 (Kirton, 1987). In studies cited by Kirton, in which the KAI scores for various occupational groups were

Copyright © 2003, Idea Group Inc. Copying or distributing in print or electronic forms without written permission of Idea Group Inc. is prohibited.

*Table 1: Ratings of Significant Levels for Non-Parametric Statistics*

| Sig. level | Statistic | Opinion(s) of p | Source | Class |
|---|---|---|---|---|
| p < 0,001 | t | "very small" | Kendall (1970) | S |
|  | r | "significant" | Kirton (1985) | P |
| p < 0,010 | t | "small" | Kendall (1970) | S |
|  | t | "significant" | Ettlie (1985) | M |
|  | r | "significant" | Kirton (1985) | P |
| p < 0,025 | w | "significant" | Weiss (1983) | I |
| p < 0,050 | t | "small" | Kendall (1970) | S |
|  | t | "significant" | Ettlie (1985) | M |
|  | r | "significant..caution.." | Olson and Ives (1981) | I |
| p > 0,100 | w | "not statistically significant" | Lusk and Kersnick (1979) | I |

r = Spearman's rank correlation co-efficient
t = Kendall's rank correlation co-efficient
w = Wilcoxon rank-sum statistic

Source: (class of literature)
I - IS,  M - Managerial,  P - Psychological,  S - Statistical

*Table 2: Qualitative Ratings Assumed for Significant Levels*

| Significance level | Qualitative ratings |
|---|---|
| p < 0,001 | Highly significant<br>Null hypothesis strongly rejected<br>Alternative hypothesis strongly supported |
| 0,001 < p < 0,050 | Significant<br>Null hypothesis rejected<br>Alternative hypothesis supported |
| 0,050 < p < 0,100 | Not very significant<br>No strong reason to reject null hypothesis<br>Weak support for alternative hypothesis<br>Inconclusive result |
| p > 0,100 | Not significant<br>No reason to reject null hypothesis<br>No support for alternative hypothesis |

Copyright © 2003, Idea Group Inc. Copying or distributing in print or electronic forms without written permission of Idea Group Inc. is prohibited.

*Table 3: The Data Samples*

---

User R-scores
Analyst KAI scores
User KAI scores
Mean KAI scores (user and analyst) *
Analyst-user KAI score differences ($\Delta$KAI) *
Absolute analyst-user KAI score differences (|KAI|) *

User opinions (7-options) of the systems analysts
in the following respects:
  their Speed of Comprehension of system requirements
  the extent to which the User's opinions were Ignored
  the Time Wasted on peripheral issues
  the tendency to remain within a Limited problem Area

User opinions (7-options) of the systems in the
following respects:
  their Accuracy of Output
  their Reliability

User opinions (7-options) of
  their Relationships with the Analysts
  the Attitudes of Analyst
  their Dissonance with the Analyst

Analysts' Ages (Years)
Users' Ages (Years)
Analysts' ages less users' ages (|Age|) *
Absolute analyst-user age differences (|Age|) *
Analysts' Lengths Of Service (LOS) (Years)
Users' Lengths Of Service (LOS) (Years)
Length-of-service differences (Analyst - User) ( LOS) *
Absolute Length-of-service differences (|LOS|) *

---

\*  Determined from corresponding data in other samples.

---

determined, the means ranged from 78.3 to 114.0, and the standard deviations from 5.8 to 26.8 (Kirton, 1987). The two KAI-score samples obtained in this study thus exhibit basic statistics that are quite within established limits. These samples also tested to be approximately normally distributed. In addition, the differences in KAI scores between users and analysts tested decidedly normal, with $p > 0.500$. The above results are consistent with (and hence support) the theory that all three of these KAI samples are random samples from approximately normal populations.

But for the item referring to speed of comprehension of the system requirements, the null hypothesis of normality in each case is rejected at a confidence level of 5%. Certain possibilities are thus implied. These are either that the populations from which the samples were drawn are not normal, or that the samples were biased (that is, not random), or that the measuring technique was not reliable. As to the first of these possibilities, (that is, non-normal parent populations), no conclusions can be drawn. The parent populations simply are not available for analysis. As far as biasing of the samples is concerned, it is difficult to see how the sampling technique

Copyright © 2003, Idea Group Inc. Copying or distributing in print or electronic forms without written permission of Idea Group Inc. is prohibited.

could have allowed this. The only parameter to be measured that the selectors of the systems in each organization knew of in advance, was that of user resistance. However, the R-score sample, which is composed of measures of user resistance by direct observation, tested satisfactorily normal. Bias for covert reasons on the part of organizational managers was thus considered unlikely. This therefore leaves the third alternative as a possibility, namely, failure of the seven-interval measuring technique. Experience with these scales confirmed this reservation. For example, the respondents at times tried to give intermediate responses, which were discouraged by the choice-presentation. Quite clearly, this would have limited tying in the data, enhancing the credibility of the association statistics.

Fortunately, reliable measures of association were still possible in the case of most of the non-normal, seven-option samples, where the responses could be meaningfully regrouped. Some sets of responses, for example, could be redefined as dichotomies or trichotomies. Somewhat more confidence is thus indicated for measures where the responses can, on some legitimate criterion, be divided into two giving approximately 50%, or three giving some 33% of the readings in each class. Effectively, this means that an approximate maximum of 17 out of 34 readings should be present for anyone of the options 1 to 7; at least a requirement fulfilled by all the samples considered below.

## The Results

The associations were first computed as Kendall's rank correlation coefficients, ta and tb. In order to measure the significance of each association $t$, the standard normal statistic for zero association, $z(t)$, was calculated. The significance of $t$ was then determined from standard normal tables (Huntsberger $et\ al$, 1973), as a tail area. The results of this procedure for each hypothesis have been summarized in Table 4. Table 5 contains a complete summary of the consequent findings.

## The Role of Cognitive Problem-Solving Styles

Hypothesis HI, which posits a significant positive correlation between the user R-scores and arithmetic difference in analyst and user KAI scores (I KAI I), tested as significant at the 0.005 significance level. In other words, there is substantially more than minimal evidence of a positive association between user resistance and analyst-user cognitive style differences. The implication is immediately obvious: to minimize user resistance; match users with analysts of similar cognitive style. However, some caution is in order. Adaptors and innovators, according to Kirton, have need of each other in many occupational situations; the adaptor to add stability

Copyright © 2003, Idea Group Inc. Copying or distributing in print or electronic forms without written permission of Idea Group Inc. is prohibited.

*Table 4: Tests of Hypothesises: Summary of Relevant Statistics*

| Hypothesis | $t_a$ | $t_b$ | $p_t$ | $r_a$ | $r_b$ | $p_r$ |
|---|---|---|---|---|---|---|
| H1 | ,2888 | ,2981 | ,008 (s**) | ,4017 | ,4031 | ,010 (s**) |
| H2(a) | ,1747 | ,1944 | ,068 (?) | ,2654 | ,2721 | ,059 (?) |
| H2(b) | ,1070 | ,1253 | ,171 (x) | ,1508 | ,1602 | ,179 (x) |
| H2(c) | -,2389 | -,3036 | ,014 (s*) | -,3539 | -,3915 | ,012 (s*) |
| H2(d) | -,1016 | -,1142 | ,192 (x) | -,1598 | -,1646 | ,171 (x) |
| H3(a) | -,3387 | -,3922 | ,002 (s**) | -,4712 | -,4916 | ,002 (s**) |
| H3(b) | -,2478 | -,2924 | ,015 (s*) | -,3518 | -,3702 | ,017 (s*) |
| H4(a) | ,0749 | ,0773 | ,264 (x) | ,0860 | ,0864 | ,309 (x) |
| H4(b) | ,0909 | ,0943 | ,224 (x) | ,1269 | ,1274 | ,233 (x) |
| H4(c) | ,1462 | ,1521 | ,111 (x) | ,1804 | ,1812 | ,149 (x) |
| H4(d) | ,1159 | ,1208 | ,166 (x) | ,1537 | ,1546 | ,187 (x) |

| | | |
|---|---|---|
| $p_t$ | p determined using either $t_a$ or $t_b$ | |
| $p_r$ | p determined using either $r_a$ or $r_b$ | |
| (s**) | "significant" | $p < 0,010$ |
| (s*) | "significant" | $p < 0,020$ |
| (s) | "significant" | $p < 0,050$ |
| (?) | "indecisive significance" | $0,050 < p < 0,100$ |
| (x) | "not significant" | $p > 0,100$ |

n = 34
N. B. All subjective assessments of p are as per Table 2

to the innovator's higher risk operations, and the innovator to motivate potentially needed changes. To match analysts and users of similar cognitive styles would deprive the system development effort of this balance. A system developed by two innovators, for example, would be expected to reach the implementation phase quickly, but with less groundwork than harder-working adaptors would have done. Hence debugging after implementation could be extensive. Two adaptors, by contrast, should take a longer time to implement the system, as they would execute the analysis and design phases more thoroughly. The debugging effort would thus be lower. However, a greater enhancing effort would be expected, since certain novel features which motivated the system's development are likely to have been overlooked.

Obviously these conjectures needed testing, and so this study initially attempted to do so. The mean KAI scores for the analyst and user were used to measure the extent to which the analyst-user dyads were either two-innovator or two-adaptor. Associations were then sought between this sample and the length of implementation time, debugging time and enhancing time. Unfortunately, these tests had to be abandoned because the data were suspect. Almost none of the analysts could make a clear distinction between enhancing and debugging times. In fact, 14

Copyright © 2003, Idea Group Inc. Copying or distributing in print or electronic forms without written permission of Idea Group Inc. is prohibited.

*Table 5: Summary of Findings*

| | Variables tested for association | Hypothesized Association | Finding |
|---|---|---|---|
| H1: | R-score, \|KAI\| | s+ | s+ ** |
| H2(a): | Δ KAI, Analyst's speed of comprehension | s+ | ?+ # |
| H2(b): | ΔKAI, Degree analyst ignored user opinion | s+ | x |
| H2(c): | ΔKAI, Time wasted by analyst on side-issues | s+ | s- * |
| H2(d): | ΔKAI, Concentration by analyst on limited area | s- | x |
| H3(a): | R-score, Accuracy of output | s- | s- ** |
| H3(b): | R-score, Reliability of system | s- | s- * |
| H4(a): | R-score, User age | s+ | x |
| H4(b): | R-score, User Length-Of-Service (LOS) | s+ | x |
| H4(c): | R-score, Absolute analyst-user age-difference | s+ | x |
| H4(d): | R-score, Absolute analyst-user LOS-difference | s+ | x |

| | | |
|---|---|---|
| s | "significant" | $p < 0.050$ |
| ? | "indecisive significance" | $0.050 < p < 0.100$ |
| x | "not significant" | $p > 0.100$ |
| + | "positive association" | ta, tb, ra, rb $> 0$ |
| - | "negative association" | ta, tb, ra, rb $< 0$ |
| ** | $p < 0.010$ | |
| * | $p < 0.020$ | |
| # | $p < 0.070$ | |

out of 34 declined to offer even the wildest of guesses. Also, only in the case of 20 out of the 34 systems were some estimates of the installation times available, since the other 14 were under development on an on-going basis. This indicates that there may well be latent disadvantages in matching users and analysts of similar cognitive styles, despite this study's inability to find any.

It is suggested, therefore, that analysts' and users' cognitive styles be matched only where either user resistance is a high-risk, high-penalty overhead, or where any of the developing, debugging or enhancing efforts is likely to have a limited impact. Examples of the first of these types of situation were noted during the research, where radical changes in state policy had forced certain organizations into corresponding computer system changes. In such cases, failure to develop and adopt the new system quickly would have meant substantial losses. It is submitted that it is worth matching analysts and users of similar cognitive styles to minimize user resistance under such conditions.

Copyright © 2003, Idea Group Inc. Copying or distributing in print or electronic forms without written permission of Idea Group Inc. is prohibited.

Up to this stage, the association between the user R-scores and the absolute KAI score differences have only been qualitatively demonstrated. Since this association tested significant at p = 0.005, a level well below the assumed permissible maximum of 0.050, an attempt to quantify the result was made. It was found that the sample statistic In (k/r), where *k* is the absolute analyst/user KAI difference and r the R-score, is approximately normally distributed with mean -0.20503 and standard deviation 0.82417. By assuming these results for the population of analyst-user dyad, Table 6 could be constructed. This enables approximate forecasts for user resistance, given the analyst and user KAI score differences.

For the sake of comparison, the raw and relative frequencies of the ratios r/k for the 34 systems researched, which fell into each confidence interval, are given as well. It will be noticed that the relative frequency of the systems in each confidence interval agrees approximately with the corresponding confidence level. This provides some heuristic evidence that the method described above for estimating the ratio r/k is valid.

At first sight, the confidence intervals given in Table 6 appear somewhat large, particularly for the higher confidences. However, forecasts based on the lower confidences may certainly be used for decision-making purposes. Additionally,

*Table 6: Confidence Intervals for Ration R/K*

| Confidence Level (%) | Lower Limit | Upper Limit | SF (Raw) (out of 34) | SF (Rel) (%) |
|---|---|---|---|---|
| 55 | .66 | 2.30 | 19 | 56 |
| 60 | .61 | 2.45 | 21 | 62 |
| 65 | .56 | 2.66 | 21 | 62 |
| 67 | .55 | 2.75 | 22 | 65 |
| 70 | .52 | 2.89 | 24 | 71 |
| 75 | .48 | 3.17 | 24 | 71 |
| 80 | .43 | 3.53 | 25 | 74 |
| 85 | .38 | 4.02 | 27 | 79 |
| 90 | .32 | 4.74 | 31 | 91 |
| 95 | .24 | 6.17 | 33 | 97 |

SF (raw) = Raw Sample Frequency
SF (rel) = Relative Sample Frequency

Copyright © 2003, Idea Group Inc. Copying or distributing in print or electronic forms without written permission of Idea Group Inc. is prohibited.

one-sided forecasts may be preferable on occasion. For example, suppose that prior to its embarking upon a joint project, an analyst-user dyad exhibits an absolute KAI score difference of 10. Then any of the following statements, based on the values given in Table 6, are acceptable:

After implementation of the system,
1)    there is a better than 50% (namely 55%) chance that the user R-score will be at least 7;
2)    there is an approximately 80% chance that the user R-score will be at least 5; and
3)    there is an approximately 80% chance that the user R-score will be no more than 35.

If user resistance constitutes a high-risk, high-penalty overhead, then one-sided forecasts based on the upper confidence limits give a safe but high "worst case," while those based on the lower confidence limits, of course, give the reverse.

A difficulty with which an organization is likely to be faced in the forecasting of user resistance on this basis is the interpretation of the R-scores. Unlike, for example, the Centigrade temperature scale, the R-score is not a measure with which people in general are familiar. Fortunately, this problem can be resolved intuitively by relating the R-scores to the numbers of complaints made in respect of each of the systems researched. The Pearson correlation coefficient of the numbers of complaints versus the corresponding R- scores was found to be 0.9126. This means that a strong, linear relationship (not merely an association) holds between these variables. Furthermore, the best-fitting regression line passes through $(0, 0)$, since zero complaints imply a zero R-score. In other words, the numbers of complaints and the R-scores are in approximate direct proportion. Based on this finding, the constant of proportionality was estimated by taking the mean of the ratios of the R-scores to the numbers of complaints. The mean was found to be 3.913. The R-score is thus approximately four times the number of distinct complaints that the user will make, in confidence, to an independent consultant, in respect of the system and/or its manner of implementation.

The strength of the association demonstrated for hypothesis HI not only facilitates a quantitative forecast of user resistance, but also supports the use of the R-score as a valid measure of user resistance. Insofar as user satisfaction and user resistance are related, the R-score is also a potential measure of user satisfaction. Of course this single study, carried out on one, comparatively small sample of systems, is insufficient to substantiate the R-score's general use in a positively

Copyright © 2003, Idea Group Inc. Copying or distributing in print or electronic forms without written permission of Idea Group Inc. is prohibited.

prescriptive manner. For instance, as previously noted, the sensitivity of the R-score to the cognitive style of the researcher is left untested in this study.

## The Ability of Adaption-Innovation Theory to Predict Aspects of the Analyst-User Interface

Following the successful testing of hypothesis HI, significant associations between analyst-user KAI score differences and aspects of the analyst-user interface were expected. These associations are represented by hypotheses H2(a) to H2(d). Of these, hypothesis H2(c) tested significant at p = 0.020 *in precise contradiction* of the original posit. A tendency was thus demonstrated for a user *not* to view an analyst who is more innovative than he is as a person who tends to waste time on side issues. It initially seems, therefore, that A-I theory failed to predict this aspect of the analyst-user interface correctly. However, a reconsideration of hypothesis H2(c) suggests another, which both agree with A-I theory and the result obtained. That is, that the user, generally being a non-systems expert, really does not know whether the analyst is wasting time on peripheral issues or not. What he rather observes in a more innovative analyst is a confident, to-the-point individual, who does not appear to waste time on peripheral issues. Unfortunately, it can be argued that this is merely an attempt to explain away an unexpected result, since the original hypothesis was a fair one, quite as soundly based on A-I theory. The ultimate conclusion in respect of this matter must then be one of caution when trying to predict specific behaviors of an analyst-user dyad directly from A-I theory.

Hypothesis H2(a) tested inconclusively significant at p = 0.070, although stronger support may have been achieved with a better measuring technique. In other words, there is some evidence that a user will find an analyst who is more innovative than he is to comprehend system requirements relatively quickly. These results may prove useful in situations where R-score/KAI testing of the persons involved is not immediately feasible or possible. In such cases, users who comment on the quick comprehension, confidence and brevity of the analyst could be suspected of belonging to analyst-user dyads in which the analyst is the more innovative. Such observations might also motivate the later administration of KAI tests or measuring of R-scores, so that a more precise analysis of user resistance can be made.

On the basis of this study, the hypotheses H2(b) and H2(d) cannot be considered as providing useful information. However, it is of interest to note a common feature. They both involved direct questioning of the user regarding his personal relationship with the analyst, implying potential negative criticism of the analyst in some way. Although in terms of A-I theory these associations should have

Copyright © 2003, Idea Group Inc. Copying or distributing in print or electronic forms without written permission of Idea Group Inc. is prohibited.

been significant, it is submitted that few of the users responded sufficiently frankly to give reliable (that is, unbiased) results.

## The Roles of System Satisfaction Factors

Hypothesis 3(a) and H3(b) tested significant at $p = 0.020$. This suggests a strong negative association between the user's perception of the accuracy and reliability of the system, and user resistance. This study thus confirms that accuracy and reliability are key factors in the issue of user resistance. Insofar as user resistance and user satisfaction are negatively associated, these results are also in accordance with the findings of Bailey and Pearson. The higher significance of the tests for H3(a) and H3(b), together with the rather low value for the correlation co-efficients for H2(a) to H2(d), imply that factors other than cognitive style differences may play some role in user resistance. However, it must be conceded that the accuracy and the reliability of the systems were recorded as seen from the point of view of the user. These assessments, it can be argued, were coloured by the cognitive styles of the user. For example, an adaptive user might well view an innovative analyst's system as a non-conservative, higher-risk tool, in line with the general adaptor's views of innovators. Consequently, he would view the system as less accurate and less reliable. The reverse is as plausible. An innovative user might view an adaptive analyst's system as too traditional, failing to encompass all the novel features that the user believes he needs. Hence, once again, the user may view the system as less accurate and less reliable. In other words, the significant associations found for H3(a) and H3(b) actually agree with the predictions of AI theory.

## The Roles of Age and Length of Service

The hypotheses H4(a) and H4(b) support the beliefs that the age and lengths of service of users are associated with user resistance. These hypotheses were both rejected at $p=0.100$. Some doubt may be argued over the result for hypothesis H4(a), since the ages of users in the sample tested somewhat skew, with $0.050>p>0.020$. However, this apparent skewness cannot of itself explain away a low, distribution-independent correlation, unless it can be shown that the sample was deliberately biased. It is difficult to see how this was possible in the light of the research design. Each system investigated was selected without reference to the user's age or length of service. This study thus rejects the beliefs that older users, or users of longer service are more resistant than others to new information systems.

Hypotheses H4(c) and H4(d) examine alternative beliefs; namely, that users who differ substantially from the analyst in terms of age or length of service are more

Copyright © 2003, Idea Group Inc. Copying or distributing in print or electronic forms without written permission of Idea Group Inc. is prohibited.

resistant than others to the analyst's new systems. All sets of readings for these hypotheses tested normal. The results showed no support for these beliefs (at p = 0.100) either, and they are also rejected by this study.

Unfair discrimination of older employees, as suggested by Rosen and Jerdee, might occur in the IS area to older users. Such discrimination could well be motivated by studies such as Bruwer's, which suggest, inter alia, that older and longer-serving personnel are most resistant to new computer systems. Attention is drawn to the fact that neither the present study nor A-I theory, nor the literature survey by Rosen and Jerdee support these beliefs. Unfair discrimination could certainly cost both individuals and organizations dearly in terms of unnecessary retrenchment and/or transfer of experienced staff.

# CONCLUSION

Subject to the postulate that the user R-score is a legitimate measure of user resistance, certain causes and impacts of user resistance have been demonstrated by this study. It has been shown that matching analysts and users of similar cognitive styles can minimize user resistance. However, care is suggested in applying such a policy indiscriminately, since there may be disadvantages in so doing despite this study's failure to show any. The high degree of association between user R-scores and absolute analyst-user KAI score differences permits approximate forecasts of the former given the latter.

Two phenomena associated with analyst-user dyads where the analyst is the more innovative, were implied by the results. The first of these is a tendency for the user to view the analyst as brief and confident, seldom wasting time on side issues. The second is (less convincingly) that the user may comment on the unexpected speed with which the analyst comprehends system requirements or specifications. It is suggested that these features be noted as diagnostic signs of the more innovative nature of the analyst, especially in situations where formal testing has not or cannot be conducted. This study finds in favor of the hypotheses that perceived accuracy and reliability of the system are associated with lower user resistance. Consequently, further insights may be gained into a given user-analyst dyad by questioning the user on the accuracy and reliability of a system. For example, the less accurate and less reliable a user considers a system, the more user resistance can be expected, and consequently (from hypothesis HI), the greater the analyst-user cognitive style difference is likely to be.

There is no support for the beliefs that the ages and lengths of service of users are associated with user resistance: or for the alternative beliefs that users who differ

Copyright © 2003, Idea Group Inc. Copying or distributing in print or electronic forms without written permission of Idea Group Inc. is prohibited.

substantially from the analyst in terms of age or length of service are more resistant than others to new systems. Since neither the present study nor A-I theory, nor the literature survey by Rosen and Jerdee support these beliefs, organizations should be alerted to the possibility of unfair discrimination against older and more experienced users.

While some discussion of user resistance exists in the literature, no direct, quantitative measure of this phenomenon had previously been attempted. This led to the development of the R-score, which is a direct measure of resistance in terms of observable complaints about the system, and which is significantly associated with the user's level of resistance to that system.

Prior to this study, neither had Adaption-innovation theory been applied to IS development, nor had the KAI instrument been replicated in the IS field. With the aid of these, it has been shown that matching users to analysts of similar cognitive style can minimize user resistance. Furthermore, by prior administration of the KAI instrument to analysts and users, approximate forecasts of user resistance are now possible. In other words, both A-I theory and the KAI instrument have been shown to be valuable tools in assessing, understanding and forecasting user resistance (see Table 6).

This study lends some support to the method used by Bailey and Pearson (1983) to measure user satisfaction and IS success, in that the two most significant factors found by them to satisfy users, were also found to be negatively associated with user resistance (measured as R-scores) in this study. These factors were system accuracy and reliability. This further suggests that low user resistance is indeed associated with high user satisfaction, confirming that resistance and satisfaction can be used as inverse, surrogate measures for one another. The user's R-score in the post-implementation phase is thus indicated as a possible measure of system success. However, more research would be required to ensure that it is not also significantly dependent upon the cognitive style of the investigator: a factor that would preclude its use as a standard measure. The speed with which the R-score can be assessed, though an interviewing technique, would make it an attractive option to the more protracted Pearson-type questionnaire for measuring system success.

# AREAS FOR FURTHER RESEARCH

As previously mentioned, there may be a dependence of the R-score on the cognitive style of the investigator since the R-score is determined as a result of his interaction with the respondent at an interview. In a further study, the effect of the

Copyright © 2003, Idea Group Inc. Copying or distributing in print or electronic forms without written permission of Idea Group Inc. is prohibited.

researcher's cognitive style on R-score evaluation needs to be investigated. Such a study might involve, for example, the administration of the R-score by several different persons to the same group of users, to investigate whether or not there were any significant differences amongst their results. If the R-score were shown to be insensitive to the nature of the investigator, it would then be usable as a system success standard as previously noted. If, however, the reverse were shown, then its use would remain limited to user resistance in the relative sense only. In other words, between systems be investigated by the same researcher, as in this study.

It remains to be shown to what extent the R-score is associated with all forms of user resistance. It would thus be of interest to construct other measures of resistance and test their associations with the R-score. There is some indication from Markus's study that complaints from users are a predominant form of user resistance. Further research that demonstrated this conclusively would make the R-score a more convincing global measure of user resistance than could be claimed in this study.

A further area for research is an analysis of the similarities and differences of the four general cognitive cases of analyst-user dyads, as given below:
1)   Analyst an adaptor, user an adaptor
2)   Analyst an adaptor, user an innovator
3)   Analyst an innovator, user an adaptor, and
4)   Analyst an innovator, user an innovator

Of particular interest are the questions of which of these is most efficient and/or effective in the short term, and which in the long term. The main hypothesis (HI) of the current study shows that user resistance will be at a minimum when the analyst and user are of like cognitive styles. This in turn suggests that dyads I) and 4) above might be more efficient and effective than the other two in the short term. However, no reliable inferences could be made regarding the effect of matching persons of similar or dissimilar cognitive styles in the long term.

# REFERENCES

Bailey, J.E. & Pearson, S.W., (1983). Development of a tool for measuring and analyzing computer user satisfaction. *Management Science, 29* (5).

Bruwer, P. (1984). A descriptive model of success for computer-based information systems. *Information and Management, 7*.

De Brabander, B. & Thiers, G. (1984). Successful information systems development in relation to situational factors that affect effective communication between MIS-users and EDP-specialists. *Management Science, 30* (2).

Copyright © 2003, Idea Group Inc. Copying or distributing in print or electronic forms without written permission of Idea Group Inc. is prohibited.

Eitlie, J.E. (1985). The impact of interorganisational manpower flows on the innovation process. *Management Science, 31* (9).

Eveland, J.D. (1977). *Implementation of Innovation in Organisations: A Process Approach.* Doctoral thesis, University of Michigan.

Hirschheim, R. & Newman, M. (1988). Information systems and user resistance: Theory and practice. *The Computer Journal, 31* (5).

Huber, G.P. (1983). Cognitive style as a basis for MIS designs: Much ado about nothing? *Management Science, 29* (5).

Huntsberger, D.V. & Billingsley, P. (1973). *Elements of Statistical Inference.* Boston: Allyn and Bacon, Inc.

Ives, B., Olson, M.H., & Baroudi, J.J. (1983). The measurement of user information satisfaction. *Communications of the ACM, 26* (10).

Keen, P.G.W. (1981). Information systems and organisational change. *Communications of the ACM, 24* (1).

Kendall, M.G. (1970). *Rank Correlation Methods.* London: Charles Griffin & Co.

Kirton, M. (1976). Adaptors and innovators: A description and measure. *Journal of Applied Psychology, 61* (5).

Kirton, M. (1980). Adaptors and innovators in organisations. *Human Relations, 33* (4).

Kirton, M. (1984). Adaptors and innovators - Why new initiatives get blocked. *Long Range Planning, 17* (2).

Kirton, M.J. (1987). *KAI Manual.* Monograph, Occupational Research Centre, Hatfield Polytechnic, Hatfield.

Kirton, M.J. (1988). *KAI Publications List.* Monograph, Occupational Research Centre, Hatfield Polytechnic, Hatfield.

Kirton, M.J. & McCarthy, R.M. (1985). Personal and group estimates of the Kirton Inventory Scores. *Psychological Reports* 57.

Lusk, E.J. & Kersnick, M., (1979). The effect of cognitive style and report format on task performance: the MIS design consequences. *Management Science, 25* (8).

Markus, M.L. (1983). Power, politics, and MIS implementation. *Communications of the ACM. 26* (6).

Olson, M.H. & Ives, B. (1981). User involvement in system design: an empirical test of alternative approaches. *Information and Management, 4.*

Rosen, B. & Jerdee, T.H. (1976). The nature of job-related age stereotypes. *Journal of Applied Psychology, 61* (2).

Rushinek, A. & Rushinek, S.F. (1986). What makes users happy? *Communications of the ACM. 29* (7).

Copyright © 2003, Idea Group Inc. Copying or distributing in print or electronic forms without written permission of Idea Group Inc. is prohibited.

Wanous, J.P. & Lawler, E.E. (1972). Measurement and meaning of job satisfaction. *Journal of Applied Psychology,  56* (2).

Woodruff, C.K. (1980). Data processing people - Are they really different? *Information and Management,  3*.

Zmud, R.W. (1983). The effectiveness of external information channels in facilitating innovation within software development groups. *MIS Quarterly:* June.

# ENDNOTE

[1]   Most of these articles are old, but no definitive works on user acceptance of e-commerce could be found in South Africa.

Copyright © 2003, Idea Group Inc. Copying or distributing in print or electronic forms without written permission of Idea Group Inc. is prohibited.

## Appendix A

### System Satisfaction Schedule

*(Abridged)*

**System Name**:.....................................................

**Organization**:....................................................

The purpose of this interview is to assess user opinion of the above computer system.
Please note that your responses will be treated in the **strictest of confidence**.
Please be as frank and as honest as possible.

### Personal User Information

1)   Age: _____ Years _____ Months

2)   Sex: _____ (M/F)

3)   User KAI: _____

4)   For how long have you been employed by your organization?

   _____ Years _____ Months

### User Problem Schedule (R-Score)

Please enumerate all the problems which you considered or heard had occurred during the implementation and/or early life of the system.

_____

_____

_____

_____

Now please rate each of these problem areas as follows:

| (7) | (6) | (5) | (4) | (3) | (2) | (1) |
|-----|-----|-----|-----|-----|-----|-----|
| a totally insoluble problem | a very serious problem | a serious problem | a rather serious problem | a significant problem | a slight problem | no real problem |

### The following were also put to the user:

How **accurate** (correct) would you say that the system's output information is?

| ☐ | ☐ | ☐ | ☐ | ☐ | ☐ | ☐ |
|-----|-----|-----|-----|-----|-----|-----|
| excellent | very good | good | rather good | satisfactory | second-rate | poor |
| *(7)* | *(6)* | *(5)* | *(4)* | *(3)* | *(2)* | *(1)* |

How **reliable** do you consider the system's output information to be?

| ☐ | ☐ | ☐ | ☐ | ☐ | ☐ | ☐ |
|-----|-----|-----|-----|-----|-----|-----|
| excellent | very good | good | rather good | satisfactory | second-rate | poor |
| *(7)* | *(6)* | *(5)* | *(4)* | *(3)* | *(2)* | *(1)* |

Comment / . . .

Copyright © 2003, Idea Group Inc. Copying or distributing in print or electronic forms without written permission of Idea Group Inc. is prohibited.

Comment on the **relationship** between yourself and the above-mentioned systems analyst/programmer.

| □ | □ | □ | □ | □ | □ | □ |
|---|---|---|---|---|---|---|
| excellent | very good | good | rather good | satisfactory | second-rate | poor |
| (7) | (6) | (5) | (4) | (3) | (2) | (1) |

Comment on the **willingness** shown by the above-mentioned systems analyst/programmer to assist you as the user.

| □ | □ | □ | □ | □ | □ | □ |
|---|---|---|---|---|---|---|
| excellent | very good | good | rather good | satisfactory | second-rate | poor |
| (7) | (6) | (5) | (4) | (3) | (2) | (1) |

How quickly did the analyst grasp the specifications of the system which you wanted him to design?

| □ | □ | □ | □ | □ | □ | □ |
|---|---|---|---|---|---|---|
| far less time than you expected | significantly less time than you expected | slightly less time than you expected | | slightly more time than you expected | significantly more time than you expected | much more time than you expected |
| (7) | (6) | (5) | (4) | (3) | (2) | (1) |

The analyst had his own ideas, and/or disagreed with your views on how the system should operate and solve problems.

| □ | □ | □ | □ | □ | □ | □ |
|---|---|---|---|---|---|---|
| strongly agree | agree | agree to an extent | | disagree to an extent | disagree | strongly disagree |
| (7) | (6) | (5) | (4) | (3) | (2) | (1) |

The analyst had a tendency to concentrate more on peripheral issues than on the problem at hand.

| □ | □ | □ | □ | □ | □ | □ |
|---|---|---|---|---|---|---|
| strongly agree | agree | agree to an extent | | disagree to an extent | disagree | strongly disagree |
| (7) | (6) | (5) | (4) | (3) | (2) | (1) |

The analyst had a tendency to concentrate on a limited problem area and failed to foresee other problems which could occur.

| □ | □ | □ | □ | □ | □ | □ |
|---|---|---|---|---|---|---|
| strongly agree | agree | agree to an extent | | disagree to an extent | disagree | strongly disagree |
| (7) | (6) | (5) | (4) | (3) | (2) | (1) |

There was a degree of dissonance (friction) between myself and the analyst at times.

| □ | □ | □ | □ | □ | □ | □ |
|---|---|---|---|---|---|---|
| strongly agree | agree | agree to an extent | | disagree to an extent | disagree | strongly disagree |
| (7) | (6) | (5) | (4) | (3) | (2) | (1) |

_Note:_ _The italicized scores above were not displayed on the original instrument form._

Copyright © 2003, Idea Group Inc. Copying or distributing in print or electronic forms without written permission of Idea Group Inc. is prohibited.

## Chapter XII

# Electronic Commerce and Data Privacy: The Impact of Privacy Concerns on Electronic Commerce Use and Regulatory Preferences

Sandra C. Henderson, Charles A. Snyder and Terry Anthony Byrd
Auburn University, USA

## ABSTRACT

*Electronic commerce (e-commerce) has had a profound effect on the way we conduct business. It has impacted economies, markets, industry structures, and the flow of products through the supply chain. Despite the phenomenal growth of e-commerce and the potential impact on the revenues of businesses, there are problems with the capabilities of this technology. Organizations are amassing huge quantities of personal data about consumers. As a result, consumers are very concerned about the protection of their personal information and they want something done about the problem. This study examined the relationships between consumer privacy concerns, actual e-commerce activity, the importance of privacy policies, and regulatory*

Copyright © 2003, Idea Group Inc. Copying or distributing in print or electronic forms without written permission of Idea Group Inc. is prohibited.

*preference. Using a model developed from existing literature and theory, an online questionnaire was developed to gauge the concerns of consumers. The results indicated that consumers are concerned about the protection of their personal information and feel that privacy policies are important. Consumers also indicated that they preferred government regulation to industry self-regulation to protect their personal information.*

Electronic commerce (e-commerce) is revolutionizing the way we conduct business. It is changing economies, markets, and industry structures; the flow of products and services through the supply chain; as well as consumer segmentation, values, and behavior (Drucker, 1999). "It is redefining commerce, transforming industries, and eliminating the constraints of time and distance. There is not a market on the face of the earth which will be ignored (PwC, 1999a)."

Despite the phenomenal growth of e-commerce, two studies recently released by the Wharton School indicated that while total spending was up, the average number of dollars spent per consumer was down. The researchers found that the concern about privacy and an unwillingness to trust online businesses with private data are the two biggest factors contributing to the decline in the dollar amount of sales per consumer (Garfinkel, 2000). According to PricewaterhouseCoopers, without privacy protection, there is no consumer confidence in e-commerce (PwC, 1999b).

This paper reports the results of a study that explored consumers' concern about the privacy of their personal information and their perceived importance of privacy policies. We also looked at consumer preferences for the alternative regulatory approaches to protecting privacy–government regulation or industry self-regulation.

In the following sections, we first discuss the existing literature on both e-commerce and privacy. We then follow with a theoretical model that links consumer attitudes toward privacy with the different regulatory options. Details of the study's methodology and results are followed by a discussion of the findings. We conclude with a discussion of the implications for both researchers and managers.

# E-COMMERCE

Definitions of e-commerce vary widely, but in general e-commerce refers to all forms of commercial transactions involving organizations and individuals that are based upon the processing and transmission of digitized data, including text, sound, and visual images (OECD, 1997; US DOC, 2000). In its most basic form, e-commerce includes technologies such as telephones, facsimile machines, auto-mated teller machines (ATMs), electronic funds transfer (EFT), and electronic data

Copyright © 2003, Idea Group Inc. Copying or distributing in print or electronic forms without written permission of Idea Group Inc. is prohibited.

interchange (EDI). More often, though, e-commerce is simply thought of as the buying and selling of goods and services through the Internet, particularly the World Wide Web (WWW). The two most common forms of e-commerce are labeled business-to-consumer (B2C) and business-to-business (B2B) e-commerce.

B2C e-commerce may be thought of as the basic type of e-commerce because it was first exploited by retail "e-businesses" such as Amazon.com, eTrade, and eBay that were created as Internet-only versions of traditional bookstores, broker-age firms, and auction houses. These e-businesses could deliver almost unlimited content on request and could react and make changes in close to real time because of the freedom from the geographic confines and costs of running actual stores (Buckley, 1999). These factors soon caused traditional "brick and mortar" stores to launch their own online stores (e.g., Barnes and Noble, Merrill Lynch, Southebys).

B2B e-commerce has many of the same advantages that hold for B2C e-commerce organizations such as the ability to increase the services they can offer their business customers. Internet technology has helped create new relationships and to streamline and augment supply-chain processes. The roles of logistic and financial intermediaries (e.g., FedEx, UPS, American Express) are expanding as these changes are occurring (Buckley, 1999).

E-commerce is not without risks or barriers. Market conditions constantly change as new competitors can easily enter the market with new business models. Customer loyalty is fleeting, as competitors are only a mouse click away. Competi-tive advantage is short-lived, as traditional barriers are rendered irrelevant by technological advances (Oracle, 1999). One risk that has received a lot of attention recently concerns the privacy and security of personally identifiable information (PII) transmitted over the Internet and stored by the organization collecting the data (Hoffman, Novak, & Chatterjee, 1993).

# PRIVACY

According to Westin (1967), privacy is "the claim of individuals, groups or institutions to determine for themselves when, how, and to what extent information about them is communicated to others." Information privacy can be thought of as "the ability of the individual to personally control information about one's self" (Stone, Gardner, Gueutal, & McClure, 1983). Personal information privacy has become one of the most important ethical issues of the information age (Culnan, 1993; Mason, 1986; Smith, 1994).

Data privacy is a major issue facing nearly every business in every country in the world. Over the past several years, many surveys have found consistently high levels of concern about privacy (Cranor, Reagle, & Ackerman, 1999; Culnan, 1993; GVU, 1998; Harris Louis and Associates & Westin, 1991, 1994, 1996; Louis Harris & Associates, Inc., 1999; Milberg, Burke, Smith, & Kallman, 1995; Smith, Milberg, & Burke, 1996). This attention to the data privacy issue has been brought about, in

Copyright © 2003, Idea Group Inc. Copying or distributing in print or electronic forms without written permission of Idea Group Inc. is prohibited.

part, by the increasing impact of information technologies (IT) such as the Internet on daily life and by recent media attention (Smith, 1993).

The European Union's (EU) Data Protection Directive also has contributed to the concern over the protection of an individual's data privacy. The directive was enacted to protect European citizens from privacy invasions. One of the major provisions of the directive prevents the transfer of personal data to countries whose privacy laws or policies do not measure up to those of the EU.

The United States is especially concerned about the implications the directive will have on trade with Europe (Santosus, 1998). The 2-year negotiations between the US Department of Commerce and the European Commission's (EC) Internal Market Directorate resulted in the EC finally accepting the US "safe harbor" principles, which are supposed to meet the test of the European directive in protecting personal data (LaRussa, 2000; Mogg, 2000).

Even with the satisfactory outcome of the EU and US negotiations, consumers and the Federal Trade Commission (FTC) have pushed and will continue to focus on the issue of data protection in the US as rumors surface that the FTC is ready to recommend new regulations to protect online privacy (Schwartz, 2000). This is a complete turnaround from their prior position that self-regulation is the method of choice.

## Online Privacy

There have been several recent studies that addressed the issue of online privacy. The FTC conducted three studies between 1998 and 2000. These studies found that most Web sites collected personal information from consumers and most had not implemented the fair information practice principles originally outlined by the 1998 study (FTC, 1998, 1999, 2000). Culnan (1999a, 1999b) conducted two studies that revealed that once again most Web sites collected personal information while only a limited percentage implemented all of the fair information practice principles.

What these studies do not show, however, is how consumers view efforts to protect their personal data on their behalf. This study examines the variables that affect consumer views of privacy, the importance of privacy policies, and whether governmental or self-regulation is more desirable.

# RESEARCH MODEL AND HYPOTHESES

The research model proposed for this study is shown in Figure 1. The model considers the interrelationships among several factors based on the literature. The following sections describe each component of the proposed model and develop associated hypotheses.

Copyright © 2003, Idea Group Inc. Copying or distributing in print or electronic forms without written permission of Idea Group Inc. is prohibited.

*Figure 1: Research Model*

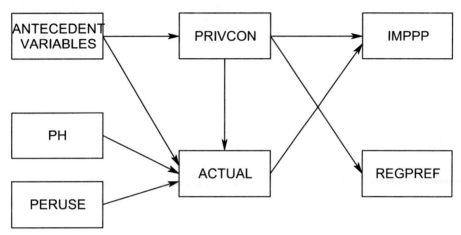

## PRIVACY CONCERNS

Privacy is a multidimensional concept. Agranoff (1991) defined information privacy in terms of three issues: data collection, data accuracy, and data confidentiality. Smith et al. (1996) identified four dimensions that constitute the construct "individual privacy concerns" (PRIVCON). The four factors are: collection, unauthorized secondary use, improper access, and errors.

The collection dimension involves concern that extensive amounts of PII are collected and stored in databases. Unauthorized secondary use deals with information that is collected for one purpose but is used for another, secondary purpose (internally within the organization collecting the PII) without authorization from the individuals. Improper access is concerned with data about individuals that are available to people without proper authorization from the individual (i.e., information sold or rented to a third party). The last area of concern involves inadequate protections against deliberate and accidental errors in personal data (Smith et al., 1996).

The results of several studies, governmental, industry and academic, show that individuals are concerned about the privacy of their personal information (Cranor et al., 1999; Culnan, 1993; Harris et al., 1991, 1994, 1996; Louis et el., 1999; Milberg et al., 1995; Milberg, Smith & Burke, 2000; Smith 1993, 1994; Smith et al., 1996; Stone et al., 1983).

## ACTUAL E-COMMERCE ACTIVITY

Prior research suggests that individuals with higher levels of concern about their information privacy may be more likely to refuse to participate in activities that require them to provide personal information (Smith et al., 1996; Stone et al., 1983). Understanding the behaviors that cause individuals to accept or reject computers

Copyright © 2003, Idea Group Inc. Copying or distributing in print or electronic forms without written permission of Idea Group Inc. is prohibited.

and systems has been one of the most challenging issues in information systems (IS) research (Swanson, 1988). Previous IS research has used intention models from social psychology as the basis for research on the determinants of user behavior (Davis, Bagozzi, & Warshaw, 1989; Swanson, 1982).

In the research model proposed for this study we did not use the behavioral intention, but we used the individuals' perceptions of their actual e-commerce activity. The research model posits that privacy concerns (attitude) have a direct impact on a consumer's actual e-commerce activities. Previous research indicates that individuals may take a variety of different actions based on their levels of concern, such as having no intention to provide personal information (Culnan, 1993; Stone & Stone, 1990). Therefore, we hypothesized:

**H1**: Higher levels of privacy concerns will result in lower levels of actual e-commerce activity.

The research model in Figure 1 also captures the effect of perceived usefulness (PERUSE) on actual e-commerce activity. The following was hypothesized:

**H2**: Higher levels of perceived usefulness result in higher levels of actual e-commerce activity.

Culnan (1993) included a variable to measure whether the respondents had shopped by mail or phone during the previous year. This study also used a variable that gauges a consumer's proclivity to purchase merchandise (PH) sight unseen. Thus the following hypothesis was proposed:

**H3**: The more likely a consumer is to purchase merchandise or services via mail, telephone or over the Internet during the past year, the higher the level of actual e-commerce use.

# IMPORTANCE OF PRIVACY POLICY

Several studies have looked at the privacy policies of Web sites (Culnan, 1999a, 1999b; FTC, 1998, 1999). In addition, the FTC issued its report, "Privacy Online: Fair Information Practices in the Electronic Marketplace," as a result of the previous studies (FTC, 2000). Neither study looked at how consumers feel about privacy policies. However, the IBM Multi-National Privacy Survey did examine the relationship between consumers and privacy policies (Louis et al., 1999). This study also examined the relationship between privacy concerns and the importance of privacy policies (IMPPP). Thus, the following hypothesis was proposed:

**H4**: The higher the level of privacy concern, the higher the perceived importance of privacy policies.

None of the previous studies have examined the impact of e-commerce usage on the perceived importance of privacy policies (IMPPP). The more PII consumers provide over the Internet, the more vulnerable their data becomes. Thus, the following was hypothesized:

Copyright © 2003, Idea Group Inc. Copying or distributing in print or electronic forms without written permission of Idea Group Inc. is prohibited.

**H5**: The higher the level of actual e-commerce activity, the more important a consumer will view an organization's privacy policy.

# REGULATORY PREFERENCE

How well corporations manage privacy often drives the desire for government regulation (Milberg et al., 2000). Several studies suggested that consumers' perceptions of how their personal data is handled by organizations impact their propensity to complain about privacy concerns and to demand governmental involvement (Harris et al., 1991, 1994, 1996; Stone & Stone, 1990). Milberg et al. (2000) found a positive direct relationship between privacy concerns and regulatory preferences (REGPREF). Thus, the following hypothesis was proposed:

**H6**: There is a positive direct relationship between consumer privacy concerns and regulatory preferences.

# ANTECEDENTS

Because life experiences tend to shape attitudes toward privacy, two antecedents were also included in the research model: (1) prior privacy invasion experience (PIE) (Culnan, 1993; Harris et al., 1991; Smith et al., 1996; Stone & Stone, 1990) and (2) technology knowledge (TK). Culnan (1993) did not find prior privacy experiences to be significant. However, Smith et al. (1996) found that persons having been the victim of personal information misuse should have stronger concerns regarding information privacy. Thus, we proposed the following hypotheses:

**H7a**: Privacy concerns are higher if an individual has experienced a prior privacy invasion.

**H7b**: An individual's actual e-commerce activity decreases if the individual has experienced a prior privacy invasion.

Prior research has established a positive relationship between experience with computing technology and a variety of outcomes such as affect towards computers and computing skill (Harrison & Rainer, 1992; Levin & Gordon, 1989). Agarwal and Prasad (1999) did not find that prior experience with similar technologies had a significant impact on attitude or behavior intentions. The Agarwal and Prasad (1999) study also did not test technology experience on actual system use. The IBM study found that groups that use computers and access the Internet to the greatest extent have relatively greater levels of confidence in companies that do business on the Internet (Louis et al., 1999). We propose that there is a significant effect of technology knowledge on the actual e-commerce activity. Thus, the following was hypothesized:

Copyright © 2003, Idea Group Inc. Copying or distributing in print or electronic forms without written permission of Idea Group Inc. is prohibited.

**H8a:** Higher levels of technology knowledge increase the privacy concerns of an individual.

**H8b:** Higher levels of technology knowledge increase the levels of actual e-commerce use.

Demographics such as gender, age, and education play a role in an individual's privacy concerns. Nickell and Pinto (1986) found that males tend to have more positive computer attitudes. Harrison and Rainer (1992) found that males exhibited significantly higher computer skill levels than females. Previous research has found evidence of a negative relationship between age and acceptance of technological change (Harrison & Rainer, 1992; Nickell & Pinto, 1986). One study posited that education is negatively related to computer anxiety (Igbaria & Parsuraman, 1989). The IBM study found that the younger, more educated, more affluent consumers are most likely to take steps to protect their privacy (Louis et al., 1999). Therefore, the following was hypothesized:

**H9a:** Gender, age, and education have a direct impact on the level of privacy concerns.

**H9b:** Gender, age, and education have a direct impact on the level of actual e-commerce activity.

# METHOD

## Sample and Procedure

The study was conducted during the spring and summer of 2000 to investigate the consumer privacy concerns and the relationships with behavioral intention to provide PII, actual e-commerce activity, and regulatory preferences. Since the research topic dealt with e-commerce over the Internet, the authors decided to put the questionnaire on the Internet. A study conducted by Brigham Young University's assessment office found that with electronic surveys, the response rate is better than with traditional mail surveys, the turnaround time is quicker, and data validity is nearly identical to mail surveys (Lindorf & Wygant, 2000).

Information on the survey was sent via email to the faculty and staff of the Department of Management at a large university in the southeast United States. In addition, a request for participants was announced in a few selected undergraduate management classes. While many criticize the use of students for business and social science research, other studies found students are adequate surrogates for decision makers (Hughes & Gibson, 1991; Remus, 1986). In this case, students are consumers who have to make decisions concerning their use of e-commerce over the Internet and providing PII to those Web sites. An unexpected group of respondents came from several participants contacting the authors for permission to give the survey URL out to people who would be interested in the study.

Copyright © 2003, Idea Group Inc. Copying or distributing in print or electronic forms without written permission of Idea Group Inc. is prohibited.

A total of 172 usable responses were submitted. Respondents were 63% men and 37% women. Sixty-two percent of the respondents were in the 18-29 age group; 26% were in the 30-49 age group; 12% were in the 50 and over age group. The respondents reported their education level as follows: 5% have a high school education or less, 42% have some college, and 53% are college graduates.

## Questionnaire Development

The questionnaire was developed from scales and techniques used in prior research. Each section was coded in hypertext markup language (HTML), JavaScript, and Active Server Pages (ASP). JavaScript was used to ensure that certain responses were completed. ASP was used to select a random Web page for the respondent to examine and to submit the responses from the completed questionnaire to the database used to collect the data.

Once the questionnaire was completed, several colleagues reviewed it to check for completeness and understandability. Several other individuals (outside academia) were asked to test the questionnaire by completing the survey online and provide comments concerning any problems. The responses confirmed that the questionnaire was clear and understandable. Internal consistency of the appropriate scales was calculated using Cronbach's alpha. The alpha for PRIVCON was 0.72, 0.88 for IMPPP, and 0.92 for TK. All the alphas were above the cutoff value suggested by Nunally (1978) of 0.70 for hypothesized measures of a construct. The questionnaire items can be found in the appendix.

## Measures

Several measures were required to evaluate the hypotheses. The following section describes each scale and the variable it measures.

**Privacy Concern**. Measurement of an  individual's privacy concern (PRIVCON) was obtained by using items taken from existing scales (Culnan, 1993; Louis et al., 1999; Smith et al., 1996). Respondents scored all items on a 5-point Likert-type scale, with Strongly Disagree to Strongly Agree.

**Actual E-Commerce Activity**. The respondent's actual e-commerce activity (ACTUAL) was obtained by using a scale adapted from a previous study (Louis et al., 1999). Respondents were asked questions concerning e-commerce activities during the past year. For each item, the respondent chose between the following responses: Never, 1-2 times, 3-5 times, 5-10 times, and Over 10 times.

**Importance of Privacy Policy**. The perceived importance of a Web Site's privacy policy (IMPPP) was measured using a scale adapted from a previous study (Louis et al., 1999). The respondents rated the importance of components of an organization's privacy policy on a 5-point Likert-type scale ranging from Not At All Important to Absolutely Essential.

**Regulatory Preference**. Regulatory preference (REGPREF) was assessed by using a similar method to Milberg et al. (2000). Respondents were asked to indicate

Copyright © 2003, Idea Group Inc. Copying or distributing in print or electronic forms without written permission of Idea Group Inc. is prohibited.

their agreement or disagreement (using a 5-point Likert-type scale, with Strongly Disagree and Strongly Agree as the endpoints) with the following questions: "the government should enact additional laws in order to protect the privacy of individuals..." and "Industries should rely on self-regulation in order to protect personal information collected..." The difference between the two measures (preference rating for governmental intervention minus preference rating for industry self-regulation) was used to indicate the degree to which the respondent preferred government regulation to industry self-regulation (Milberg et al., 2000).

**Purchasing Habits**. A consumer's propensity to purchase (PH) merchandise without the benefit of seeing and touching the item(s) before purchasing them was assessed using two items used by Culnan (1993). These items assessed whether the consumer had purchased items by mail or phone during the past year. Another item that covers Internet purchases was added. The respondents chose between the following responses: Never, 1-2 times, 3-5 times, 5-10 times, and Over 10 times.

**Perceived Usefulness**. Perceived usefulness (PERUSE) was assessed by asking the respondents how strongly they agree or disagree with a question concerning benefits of using the Internet as opposed to any potential privacy problems. Respondents rated the item on a 5-point Likert-type scale anchored by Strongly Disagree and Strongly Agree.

**Prior Privacy Invasion Experience**. Respondents were asked whether they had ever had a prior privacy invasion experience (PIE). The respondent checked either yes or no. A space was provided for an explanation of the privacy invasion experience.

**Technology Knowledge**. Technology knowledge (TK) was assessed by asking the respondent to rate the level of expertise with computer and Internet technology. For each item, the respondent chose between the following responses: Not at all knowledgeable, Somewhat knowledgeable, Knowledgeable, Very knowledgeable.

**Demographic Items**. Respondents were asked three demographic items: gender, age, and education. The range of responses for age was as follows: 18-29, 30-49, and 50+. Education was assessed using the following possible responses: High school or less, Some college, and College graduate.

# RESULTS

## Method of Analysis

EQS was used to determine the relationships among the variables shown in the path diagram shown in the model in Figure 1. EQS, a structural equation modeling software developed by Bentler (1990), implements a general mathematical and statistical approach to the analysis of linear structural equations. Using EQS, the researcher is able to check the overall goodness of fit of the proposed model and to

Copyright © 2003, Idea Group Inc. Copying or distributing in print or electronic forms without written permission of Idea Group Inc. is prohibited.

compare the relative goodness of fit of competing models, thereby assessing the need for, and strength of, different path models (Bentler, 1990; Hartwick & Barki, 1994).

## Overall Goodness of Fit

There is not one generally accepted measure of overall model goodness of fit, or even a set of optimal tests. Thus, we must rely on the use of multiple fit criteria. In this study, four goodness of fit indices were used. The first is the $\Pi^2$ statistic, which tests the proposed model against a fully saturated model–meaning that all variables are correlated (Bentler, 1990; Hartwick & Barki, 1994). A nonsignificant $\Pi^2$ value indicates good fit. However, the $\Pi^2$ is sensitive to sample size. In large samples, the $\Pi^2$ is almost always significant.

A better measure using the $\Pi^2$ statistic is to divide it by its degrees of freedom. In this case, the smaller the value, the better the fit. The literature gives several thresholds for reasonable fit: 5.0 or less (Wheaton, Muthen, Alwin, & Summers, 1977) and 3.0 or less (Carmines & McIver, 1981) and between 1.0 and 2.0 (Hair, Anderson, Tatham, & Black, 1998).

The most widely used overall goodness of fit indices are the goodness of fit index (GFI) and the adjusted goodness of fit index (AGFI). GFI measures the absolute fit of the measurement and structural models to the data. AGFI adjusts the value of the GFI to the degrees of freedom in the model. Thresholds for these indices are above 0.90 and above 0.80, respectively Chin and Todd, 1995, and Segars and Grover, 1993. A more restrictive threshold of above 0.90 for AGFI is often cited in IS research (Chin & Todd, 1995; Hair et al., 1998).

Another measure of goodness of fit is the comparative fit index (CFI), which is appropriate for all sample sizes and is thought to provide a more stable estimate than some of the other fit indices (Bentler, 1990; Hartwick & Barki, 1994). Values greater than 0.90 reflect acceptable fit.

Finally, the root mean square error of approximation (RMSEA) index measures the discrepancy in the population between the observed and estimated covariance matrices per degree of freedom. Thus, RMSEA is not affected by sample size (Garver & Mentzer, 1999). RMSEA is acceptable if the value is 0.08 or less (Hair et al., 1998).

The initial measurement model did not fit the data adequately ($\Pi^2 = 123.54$, GFI $= 0.88$, AGFI $= 0.79$, CFI $= 0.61$, RMSEA $= 0.12$) as illustrated in Table 2. Thus, the model was altered using the modification index provided by the LM test. The LM test represents the expected $\Pi^2$ decrease due to model modification (Bentler & Chou, 1993). The modifications based on this index are shown in Figure 2. The following were added to the model: covariances between some of the antecedent variables and direct paths between GENDER and IMPPP, between TK and IMPPP, and between PERUSE and REGPREF. The covariances added are supported by the significant correlations between the antecedent variables (see Table 1). Previous studies showing that gender and technology skills/knowledge have an impact on

Copyright © 2003, Idea Group Inc. Copying or distributing in print or electronic forms without written permission of Idea Group Inc. is prohibited.

computer attitudes provide support for the direct paths that were added to the model (Agarwal & Prasad, 1999; Harrison & Rainer, 1992).

The final model fit the data quite well ($\Pi^2$ = 49.71, GFI = 0.95, AGFI = 0.89, CFI = 0.91, RMSEA = 0.06), as illustrated in Table 2, giving evidence that the model is supported. Table 3 illustrates the standardized coefficients and t-values for the paths between the antecedent variables and the PRIVCON and ACTUAL variables. All the covariances between the antecedent variables were significant (see Table 4).

## Tests of Hypotheses

As with prior research, consumers are concerned about the protection of their personally identifiable information. The mean of the privacy concern measure was 3.76, where 1 = Strongly Disagree and 5 = Strongly Agree. The first four items of the measure resulted in more concern than the remaining three. The mean for the first four questions was 4.28, indicating a stronger concern with the protection of consumer personal information.

A list of the hypotheses and results can be found in Table 5. H1 was not supported (coefficient = -.056, t = .804). Higher levels of privacy concerns negatively impact actual e-commerce activity, however the results were not significantly significant. H2 was supported (coefficient = .145, t = 2.257). Support for H2 indicates that if the consumer feels that the benefits of using the Internet are sufficient, he or she will provide personal information and continue to participate in e-commerce activities despite any privacy concerns. There was strong support for H3 (coefficient = .425, t = 6.597). The results indicate that if a consumer is likely to buy goods and/or services sight unseen, it does not matter if the medium is the telephone, the mail, or the Internet.

H4 was supported (coefficient = .424, t = 6.195), indicating that as consumers are more concerned with the vulnerability of their personal information, the greater they perceive the importance of privacy policies. An additional finding from the final model shows a significant negative relationship between gender and the perceived importance of the privacy policy (coefficient = -.224, t = 3.145). The IBM study found that males place more importance on an organization's privacy policy (Louis et al., 1999). Technology knowledge was also found to impact the importance of privacy policies (coefficient = -.165, t = 2.296). H5 was not supported (coefficient = -.129, t = 1.804). Actual e-commerce usage does not significantly impact the perceived importance of an organization's privacy policy.

The findings indicate that H6 was supported (coefficient = .264, t = 3.589). As consumers become more concerned with the protection of their personal data, they become more likely to prefer governmental regulation to industry self-regulation. Another finding that was not hypothesized was the significant relationship between the actual e-commerce activity and regulatory preferences (coefficient = .153, t = 2.060). This possibly indicates that if a consumer provides personal information to a Web site, he or she would prefer stronger laws–as opposed to self-regulation–to ensure that the personal information is protected.

Copyright © 2003, Idea Group Inc. Copying or distributing in print or electronic forms without written permission of Idea Group Inc. is prohibited.

*Figure 2: Final Model*

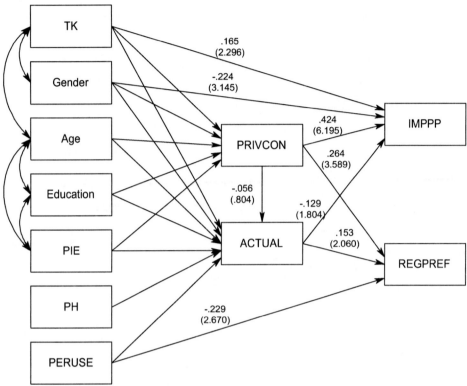

One of the two parts to H7 was supported. H7a was supported (coefficient = .333, t = 4.332), indicating that if an individual has experienced a privacy invasion experience, he or she will be more concerned with the protection of their personal data. H7b was not supported (coefficient = -.038, t = .522), indicating that even a privacy invasion experience will not keep an individual from participating in e-commerce over the Internet.

Only one of the two-part H8 was supported. H8a (coefficient = -.011, t = .141) was not supported. Increased technology knowledge did not have a significant effect on privacy concerns. However, higher levels of technology knowledge do tend to significantly increase actual e-commerce use (H8b; coefficient = .185, t = 2.681).

H9a was partially supported. Age and education had a significant effect on the level of privacy concerns (coefficient = .170, t = 2.154; coefficient = -.170, t = 2.130; respectively). Gender did not significantly impact the level of privacy concern (coefficient = .066, t = .884). H9b was also partially supported. Education does not have a significant impact on the actual level of e-commerce activity (coefficient = .107, t = 1.506). Age and gender did impact the level of actual e-commerce activity (coefficient = -.145, t = 2.035; coefficient = .166, t = 2.475; respectively). Younger

Copyright © 2003, Idea Group Inc. Copying or distributing in print or electronic forms without written permission of Idea Group Inc. is prohibited.

*Table 1:Correlation Matrix with Standard Deviations*

| | TK | GENDER | AGE | EDU | PIE | PH | PERUSE | PRIVCON | ACTUAL | IMPPP | REGPREF |
|---|---|---|---|---|---|---|---|---|---|---|---|
| TK | (1.496) | | | | | | | | | | |
| GENDER | .271** | (.485) | | | | | | | | | |
| AGE | -.220** | -.061 | (.687) | | | | | | | | |
| EDU | .110 | .046 | .276** | (.591) | | | | | | | |
| PIE | .015 | .058 | .194** | .304** | (.431) | | | | | | |
| PH | .171* | .101 | .131* | .215** | .194** | (2.184) | | | | | |
| PERUSE | .067 | .136* | .041 | .024 | -.099 | .176* | (1.110) | | | | |
| PRIVCON | -.041 | .051 | .164* | -.011 | .310** | -.014 | -.168* | (3.769) | | | |
| ACTUAL | .346** | .255* | -.113 | .171* | .029 | .480** | .265** | -.130* | (5.089) | | |
| IMPPP | .040 | -.189** | .102 | -.050 | .128* | -.146* | -.206** | .421** | -.197** | (2.574) | |
| REGPREF | -.059 | .085 | .010 | .069 | .149* | -.016 | -.222** | .285** | .061 | .212** | (1.774) |

Note: Standard deviations enclosed in parentheses.

* Correlation significant at the 0.05 level; ** Correlation significant at the 0.01 level.

Copyright © 2003, Idea Group Inc. Copying or distributing in print or electronic forms without written permission of Idea Group Inc. is prohibited.

*Table 2: Goodness-of-Fit Indices for the Initial and Final Models*

| Measure | Initial Model | Final Model |
|---|---|---|
| $\Pi^2$ (d.f.) | 123.543 (38) | 49.712 (30) |
| p-value | <0.000 | 0.013 |
| $\Pi^2$ / d.f. | 3.251 | 1.657 |
| GFI | 0.878 | 0.948 |
| AGFI | 0.789 | 0.886 |
| CFI | 0.613 | 0.912 |
| RMSEA | 0.119 | 0.064 |

*Table 3: Standardized Coefficients and T-Values for Final Model*

| | PRIVCON | ACTUAL | IMPPP | REGPREF |
|---|---|---|---|---|
| TK | -.011(.141) | .185(2.681) | .165(2.296) | |
| GENDER | .066(.884) | -.166(2.475) | -.224(3.145) | |
| AGE | .170(2.154) | -.145(2.035) | | |
| EDUCATION | -.170(2.130) | .107(1.506) | | |
| PIE | .333(4.332) | -.038(.522) | | |
| PH | | .425(6.597) | | |
| PERUSE | | .145(2.257) | | |
| PRIVCON | | -.056(.804) | .424(6.195) | .264(3.589) |
| ACTUAL | | | -.129(1.804) | .153(2.060) |

*Table 4: Coefficients and T-Values for Covariances between Antecedent Variables*

| Covariance | Coefficient | T-Value |
|---|---|---|
| TK - GENDER | .256 | 3.255 |
| TK - AGE | -.232 | 3.128 |
| AGE - EDUCATION | .284 | 3.589 |
| AGE - PIE | .205 | 2.635 |
| EDUCATION - PIE | .313 | 3.811 |

Copyright © 2003, Idea Group Inc. Copying or distributing in print or electronic forms without written permission of Idea Group Inc. is prohibited.

*Table 5: Hypotheses and Results*

| | Hypotheses | Results |
|---|---|---|
| **H1:** | Higher levels of privacy concerns will result in lower levels of actual e-commerce activity. | Not supported |
| **H2:** | Higher levels of perceived usefulness result in higher levels of actual e-commerce activity. | Supported |
| **H3:** | The more likely a consumer is to purchase merchandise or services via mail, telephone, or over the Internet during the past year, the higher the level of actual e-commerce use. | Supported |
| **H4:** | The higher the level of privacy concern, the higher the perceived importance of privacy policies. | Supported |
| **H5:** | The higher the level of actual e-commerce activity, the more important a consumer will view an organization's privacy policy. | Not supported |
| **H6:** | There is a positive direct relationship between consumer privacy concerns and regulatory preferences. | Supported |
| **H7a:** | Privacy concerns are higher if an individual has experienced a prior privacy invasion. | Supported |
| **H7b:** | An individual's actual e-commerce activity decreases if the individual has experienced a prior privacy invasion. | Not supported |
| **H8a:** | Higher levels of technology knowledge increase the privacy concerns of an individual. | Not supported |
| **H8b:** | Higher levels of technology knowledge increase the levels of actual e-commerce use. | Supported |
| **H9a:** | Gender, age, and education have a direct impact on the level of privacy concerns. | Age – Supported<br>Education – Supported<br>Gender – Not supported |
| **H9b:** | Gender, age, and education have a direct impact on the level of actual e-commerce activity. | Age – Supported<br>Education – Not supported<br>Gender - Supported |

individuals tend to use the Internet more than older individuals and males tend to use the Internet more that females.

# DISCUSSION

A specific research objective in this study was to examine the relationship between an individual's privacy concerns and e-commerce use. More specifically, the purpose was to ascertain whether privacy policies have an impact on actual e-commerce activities and whether privacy concerns impact an individual's preference for government regulation or industry self-regulation for protection of personal information. To this end, we proposed and tested a model based on

Copyright © 2003, Idea Group Inc. Copying or distributing in print or electronic forms without written permission of Idea Group Inc. is prohibited.

existing theories.

In general, the findings of the study were consistent with theoretical expectations, as most of the hypotheses were supported. The most surprising finding was that higher levels of privacy concern (H1) did not have a significant impact on actual e-commerce activity. This finding coupled with the perceived usefulness (H2) could possibly indicate that consumers will continue using the Internet if they feel the benefits and convenience of the Internet outweigh any privacy problems. In addition, consumers who have privacy invasion experiences are more concerned about privacy (H7a) issues but do not necessarily decrease actual e-commerce usage (H7b). Again, this can help explain the continued rise in e-commerce use in the face of parallel increased privacy concerns.

It is clear that individuals are concerned about the privacy of their personal information. Numerous surveys over the years have indicated a high percentage of respondents are very concerned about this issue (GVU, 1998; Harris et al., 1991, 1994, 1996; Louis et al., 1999). High levels of concern over the privacy issue have led many to express a desire for regulatory control. Our findings concerning the relationship between privacy concerns and regulatory preferences (H6) are consistent with those found by other studies (GVU, 1998; Milberg et al., 2000). The findings showed that consumers prefer government regulation to industry self-regulation with regards to the protection of their personally identifiable information.

Individuals who have higher levels of privacy concerns feel that privacy policies are important (H4). However, the negative relationship (marginal) between actual e-commerce activity and the importance of privacy policies indicates that consumers who use the Internet more place less importance on those policies. This finding suggests that high-level Internet users feel that privacy policies are either inadequate or ineffective.

## Limitations

Although previous research provides support that students are acceptable surrogates for decision makers (Hughes & Gibson, 1991; Remus, 1986), other researchers criticize the use. The use of students in this study was justified due to the subject nature and the fact that they are consumers. Since students are usually comfortable Internet users they should be suitable subjects.

## Implications and Directions for Future Research

The findings of this study have regulatory, organizational, and research implications. Now, more than ever, it appears that consumers are ready for government intervention in order to protect their personal information. This affects organizations because their efforts at self-regulation did not work well enough according to the FTC (2000). While great strides were made towards protecting the personal information of consumers, self-imposed privacy policies did not get the job

Copyright © 2003, Idea Group Inc. Copying or distributing in print or electronic forms without written permission of Idea Group Inc. is prohibited.

done (FTC, 2000). It appears that laws will be passed (Schwartz, 2000), organizations will have to make changes in order to be in compliance, and this may necessitate a change in business models for many e-commerce organizations.

There are many opportunities to develop and extend models dealing with the privacy issue. Further research could include a study to align the attitudes of consumers in relation to behavioral intentions and actual use utilizing the TRA and TAM models. Researchers could also expand the antecedent variables to gain a better understanding of the underlying issues surrounding privacy concerns, behavioral intention, and actual use of e-commerce or the Internet in general.

Another avenue of research would be to determine why consumers are indicating a preference for government regulation for protection of their personally identifiable information. Focus groups could be set up to determine why high-level Internet users feel that industry self-regulation is not as desirable as government regulation for this issue.

# CONCLUSION

Research on the issue of privacy, while not new, has recently become a hot topic. The Internet and the EU Data Protection Directive have played a large role in this interest. In addition, media attention has made the consumer very aware of the data privacy issue. They have become very aware of just how vulnerable their personal information is and they want something done about it.

Relatively speaking, e-commerce is new and not fully understood. E-commerce is growing faster than anyone expected and, of course, will experience growing pains such as the backlash stemming from the ability to invade the privacy of Internet users and the lack of data protection against any invasions. Research such as this study should help organizations understand just how critical the privacy issue is and allow adjustments to their business models to accommodate the concerns of consumers. In the long run, protecting the data privacy of consumers will help ensure success and longevity for organizations in this new economy.

# APPENDIX

**Questionnaire (adapted from the online version)**

**Instructions:** Read each question carefully and either click on the button indicating your answer or type in your answer in the space provided.

Copyright © 2003, Idea Group Inc. Copying or distributing in print or electronic forms without written permission of Idea Group Inc. is prohibited.

# Privacy Concerns

How strongly do you agree or disagree with the following statements?

|  | Strongly Disagree | Disagree | Unsure | Agree | Strongly Agree |
|---|---|---|---|---|---|
| a. Companies should never share personal information with other companies unless it has been authorized by the individuals who provided the information. | O | O | O | O | O |
| b. Companies should take more steps to make sure that unauthorized people cannot access personal information in their computers. | O | O | O | O | O |
| c. Companies are collecting too much personal information about me. | O | O | O | O | O |
| d. Consumers have lost all control over how personal information is collected and used by companies. | O | O | O | O | O |
| e. Most businesses handle the personal information they collect about customers in a proper and confidential way.* | O | O | O | O | O |
| f. Existing laws and organizational practices in the U.S. provide a reasonable level of consumer protection today.* | O | O | O | O | O |
| g. Business Web sites seem to be doing an adequate job providing notices and informing visitors how any personal information collected will be used.* | O | O | O | O | O |

# Regulatory Preferences

How strongly do you agree or disagree with the following statements?

|  | Strongly Disagree | Disagree | Unsure | Agree | Strongly Agree |
|---|---|---|---|---|---|
| a. The government should enact additional laws in order to protect the privacy of individuals using the Internet. | O | O | O | O | O |
| b. Industries should rely on self-regulation in order to protect personal information collected on Internet sites. | O | O | O | O | O |

Copyright © 2003, Idea Group Inc. Copying or distributing in print or electronic forms without written permission of Idea Group Inc. is prohibited.

## Importance of Privacy Policies

Many individuals and organizations feel that companies should adopt privacy policies. Indicate how important it is for a company to adopt and communicate or provide the following to its customers.

|  | Not At All Important | Somewhat Important | Important | Very Important | Absolutely Essential |
|---|---|---|---|---|---|
| a. An explanation to customers of what personal information is collected about them and how it will be used. | O | O | O | O | O |
| b. A choice to consumers not to have their name passed along to other companies for their marketing purposes. | O | O | O | O | O |
| c. A procedure allowing customers to see what personal information the company has stored about them and to make any needed corrections. | O | O | O | O | O |

## Internet Activity

During the past year, have you personally...

|  | Never | 1 - 2 Times | 3 - 5 Times | 5 - 10 Times | Over 10 Times |
|---|---|---|---|---|---|
| a. Visited a Web site to get information? | O | O | O | O | O |
| b. Provided personal information to a Web site such as your date of birth, your age, or your address? | O | O | O | O | O |
| c. Used a credit card to pay for information purchases on the Internet? | O | O | O | O | O |
| d. Used a credit card to pay for goods or services purchased on the Internet? | O | O | O | O | O |
| e. Conducted home banking online (i.e., checking balances or paying bills online)? | O | O | O | O | O |
| f. Invested in the stock market or traded stocks online? | O | O | O | O | O |
| g. Participated in an online auction as a seller? | O | O | O | O | O |
| h. Participated in an online auction as a buyer? | O | O | O | O | O |

Copyright © 2003, Idea Group Inc. Copying or distributing in print or electronic forms without written permission of Idea Group Inc. is prohibited.

## Purchasing Habits

During the past year, have you personally...

|  | Never | 1 - 2 Times | 3 - 5 Times | 5 - 10 Times | Over 10 Times |
|---|---|---|---|---|---|
| a. Bought something from a catalog or brochure sent to your residence? | O | O | O | O | O |
| b. Bought any product or service offered to you by a telephone call to your residence? | O | O | O | O | O |
| c. Bought any product or service through a Web site on the Internet? | O | O | O | O | O |

## Perceived Usefulness

How strongly do you agree or disagree with the following statement?

|  | Strongly Disagree | Disagree | Unsure | Agree | Strongly Agree |
|---|---|---|---|---|---|
| a. The benefits of using the Internet to get information, send e-mail, and to shop far outweigh the privacy problems that are currently being worked on. | O | O | O | O | O |

## Prior Privacy Invasion Experience

As far as you know, as a consumer...
Have you personally ever been the victim of what you felt was an improper invasion of privacy by a business or organization?    • No  • Yes

## Technology Knowledge

How knowledgeable would you say you are when it comes to...

|  | Not At All Knowledgeable | Somewhat Knowledgeable | Knowledgeable | Very Knowledgeable |
|---|---|---|---|---|
| a. Computer technology? | O | O | O | O |
| b. Internet technology? | O | O | O | O |

Copyright © 2003, Idea Group Inc. Copying or distributing in print or electronic forms without written permission of Idea Group Inc. is prohibited.

## Demographic Information

Gender       O Female
                      O Male

Age           O 18 - 29
                      O 30 - 49
                      O 50+

Education    O High school or less
                      O Some college
                      O College graduate

# REFERENCES

Agarwal, R. and Prasad, J. (1999). Are individual differences germane to the acceptance of new information technologies? *Decision Sciences*, 30(2), 361-191.

Agranoff, M. H. (1991). Controlling the threat to personal privacy: Corporate policies must be created. *Journal of Information Systems Management*, 8, 48-52.

Ajzen, I. and Fishbein, M. (1980). *Understanding Attitudes and Predicting Social Behavior*. Englewood Cliffs, NJ: Prentice-Hall.

Bentler, P. M. (1990). Comparative fit indexes in structural models. *Psychological Bulletin*, 107(2), 234-246.

Bentler, P. M. and Chou, C.-P. (1993). Some new covariance structure model improvement statistics. In Bollen, K. A. and Long, J. S. (Eds.), *Testing Structural Equation Models*. Newbury Park, CA: Sage.

Buckley, P. (1999). The emerging digital economy II, *US Department of Commerce*. Retrieved February 1, 2000 on the World Wide Web: http://www.ecommerce.gov/ede/chapter1.html.

Carmines, E. G. and McIver, J. P. (1981). Analyzing models with unobserved variables: Analysis of covariance structures. In Bohrnstedt, G. W. and Borgatta, E. F. (Eds.), *Social Measurement: Current Issues*. Newbury Park, CA: Sage.

Copyright © 2003, Idea Group Inc. Copying or distributing in print or electronic forms without written permission of Idea Group Inc. is prohibited.

Chin, W.W. and Todd, P.A. (1995). On the use, usefulness, and ease of use of structural equation modeling in MIS research: A note of caution. *MIS Quarterly*, 19(2), 237-246.

Cranor, L. F., Reagle, J. and Ackerman, M. S. (1999). Beyond concern: Understanding net users' attitudes about online privacy. *AT&T Labs-Research Technical Report TR 99.4.3*. Retrieved February 1, 2000 on the World Wide Web: http://www.research.att.com/library/TRs/99/99.4/99.4.3/report.htm.

Culnan, M. J. (1993). How did they get my name? An exploratory investigation of consumer attitudes toward secondary information use. *MIS Quarterly*, 17(3), 341-363.

Culnan, M. J. (1999a). *The Georgetown Internet Privacy Policy Survey*. Retrieved December 1, 1999 on the World Wide Web: http://www.msb.edu/faculty/culnanm/gippshome.html.

Culnan, M. J. (1999b). Privacy and the top 100 Web sites: A report to the Federal Trade Commission. *The Online Privacy Alliance*. Retrieved December 1, 1999 on the World Wide Web: http://www.msb.edu/faculty/culnanm/gippshome.html.

Davis, F. D., Bagozzi, R. P. and Warshaw, P. R. (1989). User acceptance of computer technology: A comparison of two theoretical models. *Management Science*, 35(8), 982-1003.

Drucker, P. F. (1999). Beyond the information revolution, *The Atlantic Monthly*.

Fishbein, M. and Ajzen, I. (1975). *Belief, Attitude, Intention and Behavior: An Introduction to Theory and Research*. Reading, MA: Addison-Wesley.

FTC. (1998). Privacy online: A report to congress. *Federal Trade Commission*. Retrieved on the World Wide Web: http://www.ftc.gov/privacy.

FTC. (1999). Self-regulation and privacy online: A report to Congress. *Federal Trade Commission*. Retrieved December 1, 1999 on the World Wide Web: http://www.ftc.gov/privacy.

FTC. (2000). Privacy online: Fair information practices in the electronic marketplace. *Federal Trade Commission*. Retrieved June 21, 2000 on the World Wide Web: http://www.ftc.gov/reports.

Garfinkel, S. (2000). Protecting your privacy concerns about third-party monitoring putting a damper on e-commerce. *Boston Globe*, January, C4.

Garver, M. S. and Mentzer, J. T. (1999). Logistics research methods: Employing structural equation modeling to test for construct validity. *Journal of Business Logistics*, 20(1), 33-57.

GVU. (1998). GVU's 10th WWW user survey. *Georgia Tech Graphics, Visualization & Usability Center*. Retrieved on the World Wide Web: http:/

Copyright © 2003, Idea Group Inc. Copying or distributing in print or electronic forms without written permission of Idea Group Inc. is prohibited.

/www.gvu.gatech.edu/user_surveys.

Hair, J. F., Jr., Anderson, R. E., Tatham, R. L. and Black, W. C. (1998). *Multivariate Data Analysis* (5th ed.). Englewood Cliffs, NJ: Prentice Hall.

Harris Louis and Associates and Westin, A. F. (1991). *Harris-Equifax Consumer Privacy Survey 1991.* Atlanta, Georgia: Equifax, Inc.

Harris Louis and Associates and Westin, A. F. (1994). *Equifax-Harris Consumer Privacy Survey 1994.* Atlanta, Georgia: Equifax, Inc.

Harris Louis and Associates and Westin, A. F. (1996). *The 1996 Equifax-Harris Consumer Privacy Survey.* Atlanta, Georgia: Equifax, Inc.

Harrison, A. W. and Rainer, R. K., Jr. (1992). The influence of individual differences on skill in end-user computing. *Journal of Management Information Systems, 9*(1), 93-111.

Hartwick, J. and Barki, H. (1994). Explaining the role of user participation in information system use. *Management Science, 40*(4), 440-465.

Hoffman, D. L., Novak, T. P. and Chatterjee, P. (1993). Commercial scenarios for the Web: Opportunities and challenges, *Journal of Computer-Mediated Communication, 1.*

Hughes, C. T. and Gibson, M. L. (1991). Students as surrogates for managers in a decision-making environment: An experimental study. *Journal of Management Information Systems, 8*(2), 153-166.

Igbaria, M. and Parsuraman, S. (1989). A path analytic study of individual characteristics, computer anxiety, and attitudes toward microcomputers. *Journal of Management, 15*(3), 373-388.

LaRussa, R. S. (2000, June 9). Cover letter to U.S. organizations. *Acting Under Secretary for International Trade Administration.* Retrieved on the World Wide Web: http://www.ita.doc.gov/td/ecomm/LaRussaLetJune2000.htm.

Levin, T. and Gordon, C. (1989). Effect of gender and computer experience on attitudes towards computers. *Journal of Educational Computing Research, 5*(1), 69-88.

Lindorf, R. and Wygant, S. (2000). Surveying collegiate surfers-Web methodology or mythology? *Quirks Marketing Research Review.* Retrieved June 20, 2000 on the World Wide Web: http://www.surveynetwork.com/surveys.asp.

Louis Harris & Associates Inc. (1999). *IBM Multi-National Consumer Privacy Survey.* New York, NY: IBM Global Services.

Mason, R. O. (1986). Four ethical issues of the information age. *MIS Quarterly, 10*(1), 4-12.

Milberg, S. J., Burke, S. J., Smith, H. J. and Kallman, E. A. (1995). Values, personal information privacy, and regulatory approaches. *Communications of the ACM, 38*(12), 65-74.

Copyright © 2003, Idea Group Inc. Copying or distributing in print or electronic forms without written permission of Idea Group Inc. is prohibited.

Milberg, S. J., Smith, H. J. and Burke, S. J. (2000). Information privacy: Corporate management and national regulation. *Organization Science*, 11(1), 35-57.

Mogg, J. F. (2000, July 27). Letter from Commission Services transmitting the European Commission's adequacy finding, European Commission Internal Market Director-General. Retrieved September 7, 2000 on the World Wide Web: http://www.ita.doc.gov/td/ecom/EUletter27JulyHeader.htm.

Nickell, G. S. and Pinto, J. N. (1986). The computer attitude scale. *Computers in Human Behavior*, 2, 301-306.

Nunally, J.C. (1978). *Psychometric Theory*, 2nd ed., McGraw-Hill: New York.

OECD. (1997). OECD Policy Brief No. 1-1997: Electronic commerce. *Organization for Economic Co-operation and Development*. Retrieved March 7, 2000 on the World Wide Web: http://www.oecd.org/publications/Pol_brief/9701_Pol.htm.

Oracle. (1999). The Internet changes everything. *CIO*.

PwC. (1999a). Electronic Business Outlook. *PricewaterhouseCoopers LLP*. Retrieved July 13, 2000 on the World Wide Web: http://www.e-business.pwcglobal.com/pdf/PwCPrivacy.pdf. C:\MyFiles\ECPrivacy.wpd.

PwC. (1999b). Privacy...A weak-link in the cyber-chain: Privacy risk management in the information economy. *PricewaterhouseCoopers LLP*. Retrieved July 13, 2000 on the World Wide Web: http://www.e-business.pwcglobal.com/pdf/PwCPrivacy.dbf.

Remus, W. (1986). Graduate students as surrogates for managers in experiments on business decision making. *Journal of Business Research*, 14, 19-25.

Santosus, M. (1998). Too much ado about nothing. *CIO*, August.

Schwartz, J. (2000). FTC to propose new online privacy rules. *The Washington Post*, E1.

Segars, A. H. and Grover, V. (1993). Re-examining perceived ease of use and usefulness: A confirmatory factor analysis. *MIS Quarterly*, 17(4), 517-525.

Smith, H. J. (1993). Privacy policies and practices: Inside the organizational maze. *Communications of the ACM*, 36(12), 105-122.

Smith, H. J. (1994). *Managing Privacy: Information Technology and Corporate America*. Chapel Hill, NC: University of North Carolina Press.

Smith, H. J., Milberg, S. J. and Burke, S. J. (1996). Information privacy: Measuring individuals' concerns about organizational practices. *MIS Quarterly*, 20(2), 167195.

Stone, E. F., Gardner, D. G., Gueutal, H. G. and McClure, S. (1983). A field experiment comparing information—privacy values, beliefs, and attitudes across several types of organizations. *Journal of Applied Psychology*, 68(3), 459-468.

Copyright © 2003, Idea Group Inc. Copying or distributing in print or electronic forms without written permission of Idea Group Inc. is prohibited.

Stone, E. F. and Stone, D. L. (1990). Privacy in organizations: Theoretical issues, research findings, and protection mechanisms. In Rowland, K. M. and Ferris, F. R. (Eds.), *Research in Personnel and Human Resources Management,* 349-411. Greenwich, CT: JAI Press.

Swanson, E. B. (1982). Measuring user attitudes in MIS research: A review. *OMEGA,* 10, 157-165.

Swanson, E. B. (1988). *Information System Implementation: Bridging the Gap Between Design and Utilization.* Homewood, IL: Irwin.

US DOC. (2000). Defying definition. *U.S. Department of Commerce.* Retrieved March 7, 2000 on the World Wide Web: http://www.ecommerce.gov/6.htm.

Westin, A. F. (1967). *Privacy and Freedom.* New York: Atheneum.

Wheaton, B. B., Muthen, B., Alwin, D. F. and Summers, G. F. (1977). Assessing reliability and stability in panel models. In Heise, D. R. (Ed.), *Sociological Methodology.* San Francisco: Jossey-Bass.

This chapter was previously published in the book, *Social Responsibility in the Information Age: Issues and Controversies,* edited by Gurpreet Dhillon, copyright © 2002, Idea Group Publishing.

Copyright © 2003, Idea Group Inc. Copying or distributing in print or electronic forms without written permission of Idea Group Inc. is prohibited.

Chapter XIII

# Impersonal Trust in B2B Electronic Commerce: A Process View

Paul A. Pavlou
University of Southern California, USA

## ABSTRACT

*Although the notion of impersonal trust is not new, its significance has dramatically increased with the emergence of interorganizational eCommerce. Two types of trust are usually distinguished in interfirm exchange relations– an impersonal type created by structural arrangements, and a familiarity type arising from repeated interaction. This chapter contributes to the emerging body of knowledge regarding the role of trust in B2B eCommerce, which is primarily impersonal. The nature of trust is examined, and credibility and benevolence are defined as its distinct dimensions. Impersonal trust-primarily arising from credibility-focuses on institutional structures that B2B exchanges enable through signals and incentives to facilitate interfirm relations. Following the economic, sociological and marketing literature on the sources and processes under which trust engenders, a set of three cognitive processes that generate impersonal trust is determined. Applied to B2B exchanges, four antecedents of impersonal trust are proposed to trigger these processes: accreditation, feedback, monitoring and legal bonds. In addition, impersonal trust is proposed to increase satisfaction, reduce risk, encourage anticipated continuity and promote favorable pricing. A theoretical framework is then proposed that specifies the interrelationships between the antecedents,*

Copyright © 2003, Idea Group Inc. Copying or distributing in print or electronic forms without written permission of Idea Group Inc. is prohibited.

*underlying processes and consequences of impersonal trust in B2B eCommerce. The theoretical and managerial implications of this study on B2B eCommerce are discussed, and directions for future research are proposed.*

# INTRODUCTION

The recent outbreak of electronic exchange activities, enabled primarily by the Internet, led to the emergence of B2B eCommerce. Interorganizational exchange relationships can provide a strategic source of efficiency, a competitive advantage and increased performance (Zaheer et al., 1998). A *B2B exchange* is a new form of structural platform that acts as a virtual intermediary enabling firms to conduct any-to-any online relations. As in traditional interfirm relations (Bromiley and Cummings, 1995), trust has also been considered crucial in online exchange relationships (Brynjolfsson and Smith, 2000), perhaps more given the impersonal nature of eCommerce (Keen, 2000). Trust in B2B eCommerce is mostly impersonal and it is created by structural arrangements through signals and incentives, whereas trust in traditional exchanges has been mostly based on familiarity, arising from repeated interaction. Impersonal trust is likely to be important where no social relations exist, relationships are episodic, there is information asymmetry and uncertainty, and there is some important delegation of authority between firms (Shapiro, 1987). Therefore, the context of B2B eCommerce resembles the characteristics where impersonal trust should be necessary. Hence, interfirm relations have been undergoing some dramatic changes, making the role of impersonal trust in B2B eCommerce of fundamental theoretical and managerial importance. Empirical evidence also suggests that B2B eCommerce moves away from basic transactions towards interfirm collaboration (Dai and Kauffman, 2000), making impersonal trust increasingly important. Therefore, this chapter attempts to shed light on the nature, antecedents and consequences of interorganizational trust[1] that is embedded in the impersonal context of B2B eCommerce.

Practically all transactions require an element of trust, especially those conducted in an uncertain environment. However, trust in B2B eCommerce does not comply with the traditional dyadic context of familiarity-based trust. The traditional setting for establishing trust based on familiarity not only may not be realistic in B2B eCommerce, but it could also limit its extent. Even if there is a rich tradition of scholarly research focused on familiarity-based trust in interfirm exchange relations (Geyskens et al., 1998), there is no agreed-upon understanding of interorganizational impersonal trust. In today's B2B exchanges, the traditional setting of establishing trust based on reputation, familiarity and length of the relationship (Doney and Cannon, 1997) may not be readily obtainable. In addition,

Copyright © 2003, Idea Group Inc. Copying or distributing in print or electronic forms without written permission of Idea Group Inc. is prohibited.

the absence of salespeople makes trust based on the salesperson's expertise, likeability and similarity mostly unavailable. Therefore, an impersonal type of trust may be more appropriate in B2B eCommerce. There is an urgent need to go beyond traditional dyadic relationships and examine a larger context of buyer-supplier relations in B2B eCommerce. When an increasingly large number of firms conduct business with many new, even anonymous partners, the need to understand the concept of impersonal trust in B2B eCommerce becomes fundamental.

Trust is important in impersonal exchange relationships, especially where information asymmetry and uncertainty may give rise to opportunism (Akerloff, 1970), which usually leads to mistrust, agency risks and high transaction costs. B2B eCommerce takes place in an uncertain environment that allows substantial information asymmetry between firms. Opportunism creates the problems of adverse selection and moral hazard, which are described in agency theory (Jensen and Meckling, 1976). Adverse selection occurs when firms may be motivated to misrepresent their respective abilities to the other trading firm. Moral hazard occurs when firms do no put forth the level of effort agreed upon, or fail to complete the requirements of an agreement (Mishra et al., 1998). The problems of adverse selection and moral hazard could result in excessive risk associated with online transactions, eroding the foundations of B2B eCommerce, and jeopardizing its proliferation. However, according to game theory, under suitable mechanisms opportunism does not pay off in the long run (Kandori, 1992). The institutional structures of B2B exchanges can transform many single transactions into a continuous sequence of relations between organizations, preventing opportunism through cooperative signals and incentives. Drawing from agency theory and transaction costs economics (TCE), trust may be viewed as a risk-reduction mechanism, decreasing transaction and agency costs and providing flexible trans-actions (Beccera and Gupta, 1999). B2B exchanges provide means of building trust through a series of safeguarding mechanisms, such as accreditation, feedback, monitoring and legal bonds. Since B2B exchanges are becoming an important coordination mechanism for economic activity, this chapter attempts to provide a framework to explain the process by which their mechanisms may engender impersonal trust. Moreover, the consequences of impersonal trust in B2B eCommerce are examined.

Despite the prolific differences between the traditional view of trust and impersonal trust in B2B eCommerce, this chapter proposes that impersonal trust can still complement interfirm exchange relations. The purpose of this chapter is to provide new insights into how trust develops in the impersonal context of B2B eCommerce by drawing on trust-building cognitive processes (Doney and Cannon, 1997). The global concept of trust is viewed as a two-dimensional construct in

Copyright © 2003, Idea Group Inc. Copying or distributing in print or electronic forms without written permission of Idea Group Inc. is prohibited.

terms of the dimensions of credibility and benevolence. Drawing from the literature on the sources and cognitive processes through which trust engenders, I propose that impersonal trust is associated with the dimension of credibility, and four antecedents of impersonal trust are then extracted. Furthermore, I examine how impersonal trust influences satisfaction, perceived risk, anticipated continuity and favorable pricing. More specifically, I propose a conceptual framework to describe a set of interrelationships between the antecedents and consequences of impersonal trust by attempting to answer these research questions: 1) What is the nature of impersonal trust in B2B eCommerce? 2) Which are the antecedents and consequences of impersonal trust?

The chapter is structured as follows: the next section reviews the current literature on trust, describes the nature and dimensions of impersonal trust, and portrays how a set of trust-building cognitive processes engenders credibility. A conceptual framework that examines the antecedents and consequences of impersonal trust is then developed in the context of B2B exchanges. Finally, the theoretical and managerial implications of this research are discussed in terms of the future of B2B eCommerce, and recommendations for future research are proposed.

# CONCEPTUAL DEVELOPMENT

Trust is important because it is a key element of social capital and has been related to desirable economic and social outcomes (Arrow, 1974; Geyskens et al., 1998; Zaheer et al., 1998) and a source of competitive advantage (Barney and Hansen, 1994). Trust has also been considered to reduce opportunistic behavior and transaction costs, resulting in more efficient governance (Bromiley and Cummings, 1995). Sociologists argue that buyer-supplier relations are embedded in a social context that modifies economic activity in important ways (Granovetter, 1985). For example, it can be intertwined with markets to produce "relational contracts" to ensure flexibility and opportunity (Macneil, 1980), and with hierarchies to produce "hierarchical contracts" to ensure stability and equity (Stinchcombe, 1985). In terms of theory building, trust-embedded economic theories provide a richer explanation of interfirm relationships than trust-absent theories, and also improve their descriptive and explanatory power (Beccera and Gupta, 1999). Even if rational analysis of risk can only study a calculative cooperation, independent of trust (Williamson, 1985), some authors did manage to merge economic and sociological theories and highlighted the role of trust in exchange relationships (Gulati 1995; Ouchi, 1980). In general, the role of trust is of fundamental importance and has an impact on all levels of buyer-supplier relationships.

Copyright © 2003, Idea Group Inc. Copying or distributing in print or electronic forms without written permission of Idea Group Inc. is prohibited.

There are two contextual forms of trust: impersonal created by institutional or structural arrangements through signals, incentives and rational calculation (Shapiro, 1987), or familiarity arising from long-term relationships through repeated interaction. The impersonal trust is the basis of the studies of trust from a rational cognitive perspective, often game-theoretic, mainly based on the value of keeping a reputation of honesty and competence (Dasgupta, 1988). Impersonal trust arises when no familiarity between firms is available but some structural arrangements allow subjective expectations of a firm's credibility; on the other hand, familiarity trust mainly arises from subjective anticipations of a firm's benevolence based on prior interaction. Therefore, the willingness of one firm to become vulnerable to another firm's actions depends both on the familiarity and impersonal types of trust. While there is an extensive literature on the antecedents and consequences of familiarity trust in buyer-supplier relationships (Doney and Cannon, 1997; Geyskens et al., 1998), the literature on impersonal trust is in many aspects deficient, especially at the empirical level.

Trust is formally defined as the subjective probability with which a firm assesses that another firm will perform a particular transaction according to its confident expectations in an uncertain environment. This definition captures three important attributes of trust: first, the subjective probability embraces the fact that trust is not an objective anticipation; second, the confident expectation encompasses a possibility of a (mutually) beneficial outcome; finally, the uncertain environment suggests that delegation of authority from one firm to another may have adverse (harmful) effects to the entrusting firm in case of betrayal. Therefore, trust is the subjective evaluation of the other firm's characteristics based on limited information (Beccera and Gupta, 1999). While trust could greatly improve the effectiveness of the market (Arrow, 1974), both economists and sociologists object that trust could ever become a stable coordinating mechanism because trust fails when cooperation is less profitable than cheating (Granovetter, 1985). However, given an institutional structure to encourage and safeguard cooperation, impersonal trust that is trust based on institutional arrangements through signals and incentives could perhaps be able to effectively coordinate economic activity.

## The Nature of Impersonal Trust

An impersonal analysis of trust enables studying rational and contextual cooperation, independent of familiarity trust that is usually irrational. The literature on interorganizational relations provides two general characteristics of trust: confidence or predictability in a firm's expectations about the other firm's behavior, and confidence in another firm's goodwill (Ring and Van de Ven, 1992). Moreover, trust has been viewed as the expectation that a firm can be relied on to fulfill

Copyright © 2003, Idea Group Inc. Copying or distributing in print or electronic forms without written permission of Idea Group Inc. is prohibited.

obligations (Anderson and Weitz, 1989), behave in a predictable manner, act fairly and not take unfair advantage of another firm, even given the chance (Anderson and Narus, 1990). Credibility arises from the belief that the other firm is honest and competent (Anderson and Narus, 1990), whereas benevolence arises from the belief that a firm is genuinely interested in the other firm's welfare and would seek mutual gains. Therefore, there is a broad consensus that there are two distinct dimensions of trust: credibility and benevolence (Ganesan, 1994), who investigated them independently and concluded that they did demonstrate different relationships with other variables. Credibility deals with predictability, acknowledging contracts and fulfilling the requirements of an agreement, while benevolence deals with expectations that a firm will not act opportunistically, even given the chance. Therefore, this research views two distinct trust dimensions: impersonal trust or credibility, which is based on the extent to which a firm believes that the other firm has the honesty and expertise to perform a transaction reliably, and familiarity trust or benevolence, which is based on the extent to which a firm believes that the other firm has intentions beneficial to both firms, even when new conditions without prior commitments arise.

The proposed view of trust readily corresponds to extant conceptualizations of trust. For example, Zaheer et al. (1998) viewed interfirm trust as three components–predictability, reliability and fairness. Impersonal trust encompasses predictability and reliability, while familiarity trust is equivalent to fairness. In addition, the two-dimensional view of trust is comparable to the three forms of trust defined by Sako and Helper (1998). First, contractual trust, which refers to the other firm being honest and fulfilling the explicit and implicit requirements of the contractual agreement, and second, competence trust, which pertains to whether the other firm is capable of fulfilling the contract, encompass impersonal trust. According to Sako and Helper, competence and contractual trust are often indistinguishable since contract default might be due to either dishonesty or mere inability. On the other side, goodwill trust, which relates to a firm's open commitment to take initiatives for mutual benefit while withholding from opportunistic behavior refers to familiarity trust. In sum, there is a hierarchy of trust, where fulfilling a minimal set of obligations constitutes credibility (impersonal trust), and honoring a broader set constitutes benevolence (familiarity trust).

B2B exchanges reduce the need for familiarity trust by structuring the transactional context in such a way that opportunism becomes irrational, while cooperation becomes a mutually beneficial solution. In this context, B2B exchanges compensate for the low levels of familiarity trust, which is difficult to accomplish among a great number of firms, by promoting impersonal trust based on credibility. In addition, transactional arrangements aim to predict most probable unforeseen

Copyright © 2003, Idea Group Inc. Copying or distributing in print or electronic forms without written permission of Idea Group Inc. is prohibited.

contingencies to avoid relying on another firm's benevolent motives. Hence, B2B exchanges provide a set of trust-building functions such as accreditation, feedback, monitoring and legal bonds that make opportunism irrational, thus promoting a trustworthy environment. The fact that many real-life anonymous B2B exchanges function without familiarity trust (e.g., Altra.com, Chemconnect.com) suggests that impersonal trust is at least sufficient for basic market transactions. Credibility can be regarded as the dimension of trust that governs economic activity along with the price mechanism in B2B eCommerce. Following Ganesan (1994) and the recommendations of Geyskens et al. (1998), I propose that trust has two theoretically and empirically distinct dimensions, credibility and benevolence. Impersonal trust based on credibility mostly applies to B2B eCommerce, which focuses on institutionalized structures and arrangements that B2B exchanges provide to create a stable context within which interfirm cooperation could develop.

## A Process View of Impersonal Trust

Doney and Cannon (1997) drew on several theories developed in social psychology, sociology, economics and marketing to isolate five cognitive processes through which trust engenders. These distinct processes by which trust can develop are the capability, the transference, the calculative, the intentionality, and the prediction process. These processes suggest a trust-building attempt, followed by a favorable outcome towards actually engendering trust. Therefore, these processes assume both an attempt towards developing trust, followed by a positive outcome. For example, the calculative process does not solely generate trust; the outcome of the calculation may generate trust given a favorable assessment of the calculation. In general, these five processes engender the global construct of trust. However, only the capability, transference and calculative processes are able to generate trust in an impersonal context. The processes of intentionality and prediction necessitate interaction and familiarity-learning from contact. Therefore, only capability, transference and calculative are proposed to act as mediating variables connecting the antecedents of impersonal trust in B2B eCommerce with credibility. For a more exhaustive view of all five impersonal and familiarity trust-building cognitive processes, see Pavlou (2001).

*Capability Process*

Can I trust a firm to have the competence to perform as expected? According to Sako and Helper (1998), competence is the source of capability trust, which assesses whether a firm is able to carry out its promises. Doney and Cannon (1997) argued that trust can be developed from evaluating a firm's competence. The capability process is unavoidable in all aspects of B2B exchange relations, and as

Copyright © 2003, Idea Group Inc. Copying or distributing in print or electronic forms without written permission of Idea Group Inc. is prohibited.

long as there is adequate information to perceive competence, this impersonal trust-building process can become the groundwork of a trustworthy relationship. Therefore, the capability process of engendering trust may develop in eCommerce to promote trust in a firm's credibility.

*Transference Process*

Can I trust a firm based on its performance in prior transactions with others? The institutional source of trust is associated with structural arrangements that build trust through incentive mechanisms (Shapiro, 1987). Trust can be gained based on reliable information received from a trusted network of firms, suggesting that trust can be transferred from one buyer to another, even if the trustor has no other experience. A firm may employ the cognitive transference process (Doney and Cannon, 1997) to analyze information from other firms to form its trust perceptions. Therefore, given a trustworthy network of firms, the transference process of engendering trust can become another element of impersonal trust and build trust in a firm's credibility.

*Calculative Process*

Can I trust a firm based on a calculation of its costs and benefits of cooperating? The economics literature suggests that the primary source of trust is based on a buyer's sober calculation of the other firm's cost and benefits of cheating (Dasgupta, 1988). Hence, trust involves a calculative process that a buyer assesses the potential losses compared to the short-term gains of a firm's non-cooperative behavior (Doney and Cannon, 1997). This process suggests that as long as it is irrational for a firm to cheat, it can be trusted since it is to its advantage to cooperate. Therefore, the calculative process can become another constituent of trustworthy transactions. This subjective calculation has different implications for the two dimensions of trust. While a firm's credibility can be trusted, benevolence cannot be generated based on the calculative process. According to the definition of benevolence, the other firm is expected to cooperate, even given the chance (greater benefits) of cheating; hence, the calculative process will always suggest that a firm's benevolence cannot be trusted because the benefits of cheating given the chance would always be greater. Therefore, the trust-building calculative process can build trust in a firm's credibility.

## Antecedents of Impersonal Trust in B2B eCommerce

An increasingly important application of interfirm eCommerce is the B2B exchange, which is an interorganizational information system (IOIS) through which multiple firms interact to identify and select partners, negotiate and execute

Copyright © 2003, Idea Group Inc. Copying or distributing in print or electronic forms without written permission of Idea Group Inc. is prohibited.

transactions. Most IOIS support the following market-making functions: identification, selection, execution and integration (Choudhury et al., 1998). Moreover, B2B exchanges provide some trust-building mechanisms that are proposed to act as antecedents of impersonal trust. Some antecedents can invoke multiple trust-building processes, and each antecedent represents a different method of developing impersonal trust through these basic processes.

## Accreditation

Accreditation or prequalification is defined as efforts undertaken *ex ante* to verify a firm's capability to perform as expected (Heide and John, 1990). The idea of accreditation makes sense only in a world with uncertainty and risk; in this sense, accreditation is a type of market signaling activity. Adverse selection problems can be managed by implementing qualifications processes that identify potential trading firms *ex ante* that have the skills necessary to transact in a B2B exchange (Bergen, Dutta and Walker, 1992). Accreditation may also take the form of screening by known track records; for example, e-steel.com (www.e-steel.com), a B2B exchange for trading steel, requires all potential participants to have prior trading experience and letters of reference from previous trading partners in order to register in its marketplace. Accreditation could be ascertained by a third-party B2B exchange, which may become a reliable means of characterizing firms. Accreditation triggers the capability process to assess a firm's capability to fulfill its promises, since qualification efforts can screen out incompetent firms. In this regard, accreditation is a signal that reduces adverse selection problems. Moreover, if a firm has information whether organizations are accredited, trust could be granted based on a firm's history and reputation. Accreditation is used as a surrogate of reputation for competence, which is transferable to other firms in a B2B exchange. Therefore, the transference process is also triggered by accreditation efforts. In sum, impersonal trust is associated with accreditation through inducing the capability and transference processes.

## Feedback

Research in game theory has shown that a properly designed third-party system can be an effective for assuring cooperation (Kandori, 1992). By introducing an appropriate feedback mechanism, each firm is transformed into a long-term player who conducts repeated transactions, constraining them into cooperative behavior. If there is a repeated play and an indeterminate ending point, formal analysis shows that organizations may arrive at a stable cooperative outcome (Radner, 1986). The feedback mechanism in many B2B exchanges is similar in nature to the suitable mechanism of trust presented by Lahno (1995). Given such

Copyright © 2003, Idea Group Inc. Copying or distributing in print or electronic forms without written permission of Idea Group Inc. is prohibited.

mechanism, firms are informed about other firm's past behavior and they are able to choose them. Hence, the probability of finding partners depends on their past behavior. On the basis of this dependency, only cooperative conduct pays in the long run; hence, rational firms tend to act trustworthy. This dependency engenders trust by triggering the calculative process, following a sober assessment that a firm's benefits of cheating are greater than the costs of lost transactions.

Feedback may also be regarded as a surrogate (signal) of good reputation (Pavlou and Ba, 2000), which is an important antecedent of trust in buyer-seller relationships (Anderson and Weitz, 1989). Therefore, reputable firms would have greater incentives to cooperate since they have a better feedback to protect than non-reputable firms do, and they are more likely to act ethically (Telser, 1980). Following the same argument, firms would eminently value long and unblemished history, since more organizations are more unlikely to destroy a good name to exploit a single transaction. Therefore, feedback triggers the transference process of engendering trust, where firms infer trustworthiness through feedback from other firms participating in a B2B exchange. Feedback mechanisms provide both signals of past experience, and also incentives for cooperation. Consequently, feedback is associated with impersonal trust by triggering both the calculative and transference processes.

### Monitoring

In B2B exchanges, monitoring may have two aspects. First, a third-party authority monitors all interfirm transactions and assures that everything is performed in accordance with the agreed terms. In case of a problem, a neutral authority attempts to solve the issue to the satisfaction of both firms, or in accordance with the prearranged agreement. Second, a third-party authority can assure that the quality of all products exchanged is in agreement with the preapproved specifications. For example, independent contractors offer quality-assurance services to the B2B exchange of Chemconnect.com (www.chemconnect.com). Therefore, agency risks of moral hazard are minimized by monitoring that discourages opportunistic behavior. B2B exchanges may continually monitor the trading activity, convey sanctions to inappropriate trading behavior and punish any wrongdoing. Third-party monitoring provides the incentives for firms to engage in cooperative and honest practices. Therefore, the calculative process suggests that trust can be built when a B2B exchange monitors the transaction if the costs of opportunistic behavior will be higher than the benefits from cheating. Given proper government, monitoring provides the incentives for firms to engage in cooperative practices since simple calculation would suggest that the costs of opportunistic behavior would exceed

Copyright © 2003, Idea Group Inc. Copying or distributing in print or electronic forms without written permission of Idea Group Inc. is prohibited.

potential short-term benefits. Hence, the calculative process can be invoked by monitoring, which is proposed to be associated with impersonal trust.

*Legal Bonds*

Written contracts are also proposed as a mechanism to reduce opportunistic behavior and moral hazard. However, contracts are only partial safeguards against opportunism since they are almost always incomplete due to unforeseen circumstances, since firms are considered boundedly rational and cannot foresee all possible states of nature (Williamson, 1985). Nevertheless, impersonal trust based on legal bonds can be built on the basis of the calculative process since it is rational to cooperate given a legal contract that increases the costs of opportunism. Therefore, legal bonds provide protection and can promote trust in a firm's credibility.

## Consequences of Impersonal Trust

*Satisfaction*

According to Anderson and Narus (1990), satisfaction is conceptualized as a very important consequence of exchange relationships, showing that satisfaction is an outcome of trust-based relationships. Mutual trust indicates equity in the exchange and promotes satisfaction. Moreover, trust enhances channel member satisfaction by reducing conflict (Anderson and Narus, 1990; Geyskens et al., 1998). In summary, satisfaction represents an important outcome of business exchange relations and a global evaluation of fulfillment exchanges relationships (Dwyer et al., 1987), in which both dimensions of trust should contribute. Following Ganesan (1994), there should be a positive relationship between satisfaction and impersonal trust.

*Perceived Risk*

Most buyer-supplier relationships are characterized by information asymmetry since one firm usually possesses uneven information regarding the transaction compared to the other firm (Mishra et al., 1998). The general problem faced by organizations is the inability to foresee and control the actions of the other firm, leading to delegation of some authority to the other party. This problem creates a double-sided agency relationship between the buyer and the supplier (Jensen and Meckling, 1976). According to Shapiro (1987), agency relationships are present in all types of social relationships from simply familiarity interactions to complex forms of a firm. Although risk is inevitable in every transaction, trust reduces the expectations of opportunistic behavior (Sako and Helper, 1998) and risk perceptions (Ganesan, 1994). Trust has been shown to reduce the perceived risk of being

Copyright © 2003, Idea Group Inc. Copying or distributing in print or electronic forms without written permission of Idea Group Inc. is prohibited.

taken advantage of from the other firm (Anderson and Weitz, 1989) and improves favorable impressions for the other firm (Anderson and Narus, 1990). Since signals and incentives were shown to build trust and reduce fears of moral hazard and adverse selection, trust should also reduce perceived risks. Consequently, trust in a firm's credibility should diminish risk perceptions, predicting a negative relationship between impersonal trust and perceived risk.

### Anticipated Continuity

Anticipated continuity is defined as the perception of a firm's expectation of future transactions in a B2B exchange. There is significant evidence to suggest a strong association between trust and a propensity to continue a relationship (Morgan and Hunt, 1994). According to Ganesan (1994) trust is a necessary ingredient for long-term orientation because it shifts the focus to future conditions. Similarly, Morgan and Hunt found a negative relationship between trust and propensity to leave, and also Anderson and Weitz (1989) showed that trust is key to maintaining continuity in buyer-supplier relationships. Therefore, trust should be associated with a firm's intention to continue participating in a B2B exchange. Firms participating in impersonal B2B exchanges usually make decisions based on objective, calculative evidence (credibility), rather than subjective evaluations (benevolence) since familiarity trust is rarely present. Nevertheless, anticipated continuity should be affected by trust in a firm's credibility, predict a positive relationship between impersonal trust and anticipated continuity.

### Pricing

A major reason for the existence of different prices is the need to compensate some firms for reducing agency risks and transactions costs (Rao and Monroe, 1996). Therefore, in an efficient or dynamic pricing mechanism, firms need to reward reputable firms with better prices to assure safe transactions, since reputable firms are more likely to reduce transaction and agency costs. Similarly, in B2B exchanges, trustworthy firms are likely to reduce such costs and receive more favorable pricing. This phenomenon could be explained by the notion of returns to reputation (Shapiro, 1983), where reputable agents tend to receive more favorable terms. On the contrary, organizations tend to mandate compensation for the risk they are exposed to when they transact with less reputable firms. Consequently, differences in trust may cause different prices given a dynamic pricing scheme. Pavlou and Ba (2000) empirically showed that differences in trust perceptions affect price premiums and discounts in eCommerce auctions. Similarly, in B2B exchanges with dynamic pricing schemes, impersonal trust is viewed as a risk-reduction mechanism, allowing trustworthy firms to obtain more favorable

Copyright © 2003, Idea Group Inc. Copying or distributing in print or electronic forms without written permission of Idea Group Inc. is prohibited.

*Figure 1: Conceptual Framework*

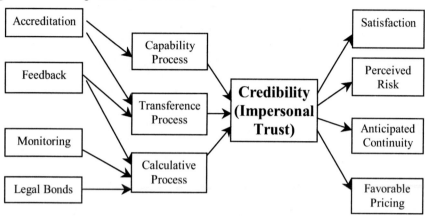

pricing terms for reducing transaction-specific risks. The complete set of antecedents, trust-building processes and consequences of impersonal trust is shown in Figure 1.

# DISCUSSION

The primary contribution of this research is that a set of interrelationships between constructs that tend to be associated with impersonal trust in B2B eCommerce are specified. First, the nature and dimensions of trust in eCommerce are described, and it is proposed that impersonal trust is mostly applicable in this context, primarily arising from the dimension of credibility. Second, the conceptual development provides the trust-building cognitive processes that may engender impersonal trust. Third, four trust-building mechanisms usually present in B2B exchanges are proposed to act as antecedents of impersonal trust by triggering the trust-building cognitive processes. Finally, four consequences of impersonal trust on key aspects of interfirm exchange relations are proposed. Although this study does not empirically examine the interrelationships among these variables, the theoretical foundation provides a comprehensive view of the role of impersonal trust in B2B eCommerce; an empirical validation of the proposed framework can be seen in Pavlou (2001). Another contribution of this research is the examination of the two distinct dimensions of trust—independently, proposing that each dimension would have different underlying cognitive processes, antecedents and consequences. While the extant literature paid particular attention to benevolence as the most important aspect of interfirm trust and viewed trust as a global construct, this

Copyright © 2003, Idea Group Inc. Copying or distributing in print or electronic forms without written permission of Idea Group Inc. is prohibited.

research showed that in B2B eCommerce, impersonal trust arising from the dimension of credibility could be studied as an independent concept in its own right. Viewing trust as a unidimensional construct may be a valid simplification when analyzing long-term relationships (Doney and Cannon, 1997). However, given the impersonal context of B2B eCommerce, this simplification underscores the importance of credibility on vital outcomes of interfirm exchange relations (Ganesan, 1994).

*Key Findings*

This research is one of the first to address the importance of impersonal trust in B2B eCommerce. In online interfirm exchange relations where familiarity trust in not readily applicable, impersonal trust can still have beneficial outcomes, primarily based on the dimension of credibility. The two dimensions of trust-credibility and benevolence–are related constructs, although theoretically distinctive from each other. While there is rich scholarly literature on the antecedents of familiarity trust, no research has investigated the effect of an extensive set of antecedents of impersonal trust. Not only does this study propose some precursors of impersonal trust in B2B eCommerce, these antecedents are theoretically grounded in the cognitive processes by which trust engenders. Whereas it may appear that impersonal processes-competence, transference and calculative-are only weak complements of the familiarity-based processes-intentionality and prediction, the proposed impersonal processes seem to be critical in establishing trust and generating favorable social and economic outcomes. Therefore, even if the notion of studying impersonal trust independently is novel, its antecedents and consequences are established based on solid ground.

*Theoretical Implications*

This research proposes that firms tend to make many decisions based on objective evidence of competence, reliability and honesty (credibility). This finding is in line with Williamson (1993) who argued that familiarity trust only exists at the individual level, whereas business relations require institutional safeguards against opportunism. Also, Ganesan (1994) showed that even in long-term interfirm relationships, credibility was the sole determinant of long-term orientation. Therefore, these findings call for reconceptualization of the role of interfirm trust and more exhaustive research on the antecedents and consequences of the impersonal concept of trust. Nevertheless, whereas the dimension of credibility is proposed to induce favorable outcomes in interfirm exchange relations, it is not the purpose of this chapter to undermine the importance of familiarity trust. However, while other researchers argued that benevolence is the only stable form of trust (Granovetter,

Copyright © 2003, Idea Group Inc. Copying or distributing in print or electronic forms without written permission of Idea Group Inc. is prohibited.

1985), I argue that credibility is also a robust form that is increasingly important in the impersonal environment of B2B eCommerce.

Many authors argued that trust is only embedded in repetitive transactions and ongoing relationships, essentially arguing that impersonal trust is not a true form of trust but a functional substitute for it (Granovetter, 1985; Williamson, 1993). Sitkin and Roth (1993) argued that 'legalistic remedies' are not effective in creating trust, while Granovetter (1985) maintained that institutional processes are 'functional substitutes' of trust. On the contrary, Shapiro (1987) argued that institutional practices and norms could provide a very strong level of trust. In sum, the role of impersonal trust in the literature has been controversial; however, given the importance of impersonal trust in today's B2B eCommerce, this chapter attempts to disentangle the complex notion of trust and encourage research on the antecedents, underlying trust-building cognitive processes and consequences of each distinct dimension. By examining the two types of trust independently, research could validly conclude which type is robust, easy to build and maintain, and consequential.

Lewicki and Bunker (1995) argued that there are different levels of trusting relationships with impersonal processes as the most fragile and familiarity ones as the most robust. However, the impersonal processes are fragile if the underlying institutional structure is fragile, and the signals and incentives are weak. Given strong and well-accepted institutional rules, practices and norms to guide B2B eCommerce, impersonal trust can also become a robust mechanism to govern interfirm economic activity. Even if familiarity trust is indeed more robust, familiarity may not always be possible in eCommerce, since it may be physically or socially difficult to personalize all interfirm relationships, and it might also have negative economic consequences (Helper, 1991). Therefore, impersonal trust can be encouraged in B2B exchanges without significant costs, and provide substantial benefits to the participating firms without forcing them into repetitive transactions.

*Managerial Implications*

This study has practical implications for the ways B2B exchanges might increase the general level of interorganizational trust. Heightening the extent of firm accreditation, improving feedback mechanisms, and ensuring an effective monitoring and legal system promotes a trustworthy environment. Failure to provide these antecedents of trust might reduce the general level of credibility in the marketplace, which would eventually force firms to seek other alternatives. Moreover, this study proposes that impersonal trust is an important determinant of satisfaction, reduced perceived risks, anticipated continuity and favorable pricing. These associations set new standards for B2B exchanges; an important consideration should be to

Copyright © 2003, Idea Group Inc. Copying or distributing in print or electronic forms without written permission of Idea Group Inc. is prohibited.

develop the appropriate mechanisms to build and sustain interfirm trust. Since the future of most B2B exchanges relies on many participating firms, high liquidity and trade volume, impersonal trust would probably become an important determinant of the future of many B2B exchanges. This research not only proposes the antecedents of impersonal trust, but it also indicates the cognitive sequences by which trust develops.

In considering their participation in B2B exchanges, firms should appreciate the role of impersonal trust that is based on functional mechanisms, institutional structures and regulations instituted in the marketplace. Since these structures provide the secure context in which interfirm exchange relations can develop, companies should either depend on these structures to trust other firms, or rely on exchange relations with familiar partners. Managers should decide which exchange relations should be based on impersonal trust, and which on familiarity trust. For more extensive managerial recommendations on how the proposed framework could be useful, see Pavlou (2000).

### Limitations and Suggestions for Future Research

This research attempts to make theoretical contributions to the academic and managerial literature on the role of impersonal trust in B2B eCommerce. The purpose is to stimulate empirical research in the area towards validating the proposed framework and shedding some light on the pragmatic nature of impersonal trust in B2B eCommerce. It should be clear that this framework proposes only a subset of the many possible relationships between trust and its antecedents, cognitive processes, consequences and other moderating variables. Hence, other important constructs could have been neglected. Future research should take a more extensive approach to cover other variables related to impersonal trust in B2B eCommerce. Therefore, future research should identify other factors that complement the proposed conceptual model, and empirically test a more complete framework.

The proposed framework may not be generalizable to other dissimilar cultures. Governance by familiarity trust is prevalent in Japanese markets; hence, impersonal B2B eCommerce might be an infeasible solution in this culture. For example, Sako (1992) compared Japanese and British companies and showed that Japanese firms exhibit higher levels of familiarity trust towards their trading partners. Therefore, the notion of worldwide B2B eCommerce may be restricted by the cultural norms of some nations that rely on familiarity relations. Future research should examine the role of ethnic culture in interfirm relations and predict the boundaries of global B2B exchanges.

Copyright © 2003, Idea Group Inc. Copying or distributing in print or electronic forms without written permission of Idea Group Inc. is prohibited.

# REFERENCES

Akerlof, G. (1970). The market for 'lemons': Quality under uncertainty and the market mechanism. *Quarterly Journal of Economics*, August (84), 488-500.

Anderson, E. and Weitz, B. (1989). Determinants of continuity in conventional firm working partnership. *Journal of Marketing*, 54(1), 42-58.

Arrow, K. J. (1974). *The Limits of Firms*, New York: Norton.

Barney, J. B. and Hansen, M. H. (1994). Trustworthiness as a source of competitive advantage. *Strategic Management Journal*, Special Issue (15), 175-190.

Beccera, M. and Gupta, A. K. (1999). Trust within the firm: Integrating the trust literature with agency theory and transaction cost economics. *Public Administration Quarterly*, 177-203.

Bergen, M. E., Dutta, S. and Walker, Jr., O. C. (1992). Agency relationships as marketing: A review of the implications and applications of agency-related theories. *Journal of Marketing*, 56, 1-24.

Bromiley, P. and Cummings, L. L. (1995). Transaction costs in firms with trust. In Bies, R., Sheppard, B. and Lewicki, R. (Eds.), *Research on Negotiation in Firms*. Greenwich, CT: JAI Press.

Brynjolfsson, E. and Smith, M. D. (2000). Frictionless commerce? A comparison of Internet and conventional retailers. *Management Science*, 46(4), 563-585.

Choudhury V., Hartzel, K. S. and Konsynski, B. R. (1998). Uses and consequences of electronic exchanges: An empirical investigation in the aircraft parts industry. *MIS Quarterly*, 471-507.

Dai, Q. and Kauffman, R. J. (2000). Business models for Internet-based eProcurement systems and B2B electronic markets: An exploratory assessment. *34th Hawaii International Conference on Systems Science*, January 2001.

Dasgupta, P. (1988). Trust as a commodity. In Gambetta, D. (Ed.), *Trust: Making and Breaking Cooperative Relations*. New York: Basil Blackwell, Inc.

Doney, P. M. and Cannon, J. P. (1997). An examination of the nature of trust in buyer-seller relationships. *Journal of Marketing*, April (61), 35-51.

Dwyer, F. R., Schurr, P. J. and Oh, S. (1987). Developing buyer-seller relationships. *Journal of Marketing*, 52(1), 21-34.

Ganesan, S. (1994), Determinants of long-term orientation in buyer-seller relationships. *Journal of Marketing*, 58, 1-19.

Geyskens, I., Steenkamp, J. B. and Kumar, N. (1998). Generalizations about trust in marketing channel relationships using meta-analysis. *International Jour-*

Copyright © 2003, Idea Group Inc. Copying or distributing in print or electronic forms without written permission of Idea Group Inc. is prohibited.

*nal in Marketing*, 15, 223-248.

Granovetter, M. (1985). Economic action and social structure: The problem of embeddedness. *American Journal of Sociology*, 91(3), 481-510.

Gulati, R. (1995). Does familiarity breed trust? The implications of repeated ties for contractual choice of alliances. *Academy of Management Journal*, 38, 85-112.

Heide, J. B. and John, G. (1990). Alliances in industrial purchasing, the determinants of joint action in buyer-supplier relationships. *Journal of Marketing Research*, 37, 24-36.

Helper, S. (1991). How much has really changed between U.S. automakers and their suppliers? *Sloan Management Review*, Summer (32), 15-28.

Jensen, M. C. and Meckling, W. H. (1976). Theory of the firm: Managerial behavior, agency costs and ownership structure. *Journal of Financial Economics*, 3, 305-360.

Kandori, M. (1992). Social norms and community enforcement. *Review of Economic Studies*, 59, 63-80.

Keen, P. G. W. (2000). Ensuring eTrust. *Computerworld*, March, 34(11), 13, 46.

Lahno, B. (1995). Trust, reputation, and exit in exchange relationships. *Journal of Conflict Resolution*, 39(3), 495-510.

Lewicki, R. J. and Bunker, B. B. (1995). Trust in relationships: A model of development and decline. In Bunker, B. B. and Rubin, J. Z. (Eds.), *Conflict, Cooperation and Justice*. San Francisco: Jossey-Bass.

Macneil, I. R. (1980). *The New Social Contract*. New Haven, CT: Yale University Press.

Mishra, D. P., Heide, J. B. and Cort, S. G. (1998). Information asymmetry and levels of agency relationships. *Journal of Marketing Research*, 35, 277-295.

Morgan, R. M. and Hunt, S. D. (1994). The commitment-trust theory of relationship marketing. *Journal of Marketing*, July (58), 20-38.

Ouchi, W. G. (1980). Market, bureaucracies and clans. *Administrative Science Quarterly*, 25, 129-141.

Pavlou, P. A. (2001). The role of trust in electronic commerce: Evidence from business-to-business and business-to-consumer electronic intermediaries. *Working Paper*, Marshall School of Business, University of Southern California.

Pavlou, P. A. (2000). Building trust in B2B relationships through electronic commerce intermediaries. *Working Paper*, Marshall School of Business, University of Southern California.

Copyright © 2003, Idea Group Inc. Copying or distributing in print or electronic forms without written permission of Idea Group Inc. is prohibited.

Pavlou, P. A. and Ba, S. (2000). Does online reputation matter? An empirical investigation of reputation and trust in online auction markets. *Proceedings of the 6th Americas Conference in Information Systems*, Long Beach, CA.

Radner, R. (1986). Repeated partnership games with imperfect monitoring and no discounting. *Review of Economic Studies*, 111, 43-57.

Ring, P. S. and Van de Ven, A. H. (1992). Structuring cooperative relationships between firms. *Strategic Management Journal*, 13, 483-498.

Sako, M. (1992). *Trust in Exchange Relations*. Cambridge, MA: University Press.

Sako, M. and Helper, S. (1998). Determinants of trust in supplier relations: Evidence from the automotive industry in Japan and the United States. *Journal of Economic Behavior and Firm*, 34, 387-417.

Shapiro, S. P. (1987). The social control of impersonal trust. *American Journal of Sociology*, 93, 623-658.

Sitkin S. B. and Roth, N. L. (1993). Explaining the limited effectiveness of 'legalistic remedies' for trust/distrust. *Firm Science*, 4, 367-392.

Stinchcombe, A. L. (1985). Stratification and firms. *Selected Papers*, Cambridge, MA: Cambridge University Press.

Telser, L. G. (1980). A theory of self-enforcing contracts. *Journal of Business*, 53(1), 27-44.

Williamson, O. E. (1975). *Exchanges and Hierarchies: Analysis and Antitrust Implications*. New York: The Free Press.

Williamson, O. E. (1985). *The Economic Institutions of Capitalism*. New York: The Free Press.

Williamson, O. E. (1993). Calculativeness, trust and economic firm. *Journal of Law and Economics*, 26, 453-486.

Zaheer, A., McEvily, B. and Perrone, V. (1998). Does trust matter? Exploring the effects of interfirm and interpersonal trust on performance. *Organization Science*, 9(2), 141-159.

This chapter was previously published in the book, *Business to Business Electronic Commerce: Challenges & Solutions*, edited by Merrill Warkentin, copyright © 2002, Idea Group Publishing.

Copyright © 2003, Idea Group Inc. Copying or distributing in print or electronic forms without written permission of Idea Group Inc. is prohibited.

# About the Authors

**Sam Lubbe** has spent more than 18 years working in the field of Information Systems. He has served some time as a user and as an IT professional. He always ensured that the organizations obtain the maximum benefit for the amount of money they invested in e-commerce. Recently, he has specialized in the area of e-commerce in SMEs. He has published some articles in this subject area. He holds a B.Com., B.Com. (Hons.), M.Com. and Ph.D. and is a Head of Department in IT at Cape Technikon in South Africa.

*   *   *

**Udo Averweg** is employed as an Information Analyst by eThekwini Municipality, Durban, South Africa. He entered the Information Technology (IT) industry during 1979 and holds a master's degree in Information Technology (*cum laude*). He is a professional member of the Computer Society of South Africa and has delivered research papers at local and international (Hawaii, Australia and Egypt) IT conferences.

**Terry Anthony Byrd** is associate professor of MIS in the Department of Management at the College of Business, Auburn University. He holds a BS in electrical engineering from the University of Massachusetts at Amherst and a PhD in management information systems from the University of South Carolina. His research has appeared in *MIS Quarterly, Journal of Management Information Systems, Decision Sciences, OMEGA, Interfaces* and other leading journals. His current research interests include the strategic management of information technology, information technology architecture and infrastructure, electronic commerce, and information technology implementation.

Copyright © 2003, Idea Group Inc. Copying or distributing in print or electronic forms without written permission of Idea Group Inc. is prohibited.

**Eric Cloete** B.Sc. (Mathematics), B.Sc. Hons. (Applied Mathematics), NHOD, M.Sc. (Computer Science), D.Tech (IT). Dr. Cloete is a born Capetonian, with a firm belief that we can take the distance out of our education with the proper use of technology. He worked for many years in tertiary education at a number of institutions in South Africa and abroad. He is currently the convener for Electronic Commerce, a final-year module in Information Systems at the University of Cape Town. His research interests include e-business, e-learning and super-computers. He is also a keen musician and plays a number of instruments on a semi-professional basis.

**Geoff Erwin** has spent more than 20 years working in the field of corporate computers and information systems. He has worked with computers as an IS professional and as a user. In all these capacities, he has been concerned with obtaining value for money from the information systems effort. He is a Dean in the Faculty of Business Informatics, Cape Technikon, South Africa.

**Rick Gibson** has more than 20 years of software engineering experience and is authorized by the Software Engineering Institute to lead assessments of software and systems development organizations. In this role, he has extensive domestic and international experience in the conduct of evaluations and the subsequent development of process maturity improvement action plans for software organizations. He is currently the department chair and an associate professor at American University, Washington, D.C., for the Department of Computer Science and Information Systems. His responsibilities, as a faculty member, include teaching graduate courses in software engineering, decision analysis, and knowledge management. He has published a variety of books, book chapters, and journal articles on software development and quality assurance.

**Jatinder N. D. Gupta** is currently Eminent Scholar of Management, Professor of Management Information Systems, and Chairperson of the Department of Accounting and Information Systems in the College of Administrative Science at the University of Alabama in Huntsville, Huntsville, Alabama. Most recently, he was Professor of Management, Information and Communication Sciences, and Industry and Technology at the Ball State University, Muncie, Indiana. He holds a PhD in Industrial Engineering (with specialization in Production Management and Information Systems) from Texas Tech University. Co-author of a textbook in Operations Research, Dr. Gupta serves on the editorial boards of several national and international journals. Recipient of the Outstanding Faculty and Outstanding Researcher awards from Ball State University, he has published numerous papers

Copyright © 2003, Idea Group Inc. Copying or distributing in print or electronic forms without written permission of Idea Group Inc. is prohibited.

in such journals as *Journal of Management Information Systems, International Journal of Information Management, INFORMS Journal of Computing, Annals of Operations Research*, and *Mathematics of Operations Research*. More recently, he served as a co-editor of a special issue on *Neural Networks in Business* of *Computers and Operations Research* and a book entitled, *Neural Networks in Business: Techniques and Applications*. His current research interests include information and decision technologies, scheduling, planning and control, organizational learning and effectiveness, systems education, and knowledge management. Dr. Gupta is a member of several academic and professional societies including the Production and Operations Management Society (POMS), the Decision Sciences Institute (DSI), and the Information Resources Management Association (IRMA).

**Sandra C. Henderson** is a doctoral candidate of MIS in the Department of Management at Auburn University. She holds a master's of accountancy with a concentration in accounting information systems from Florida State University and a BS in Accounting from Albany State University. Her research has appeared in *Information & Management* and several proceedings. Her research interests include information privacy, electronic commerce, and systems development.

**Mike Mullany** was born in Pietermaritzburg, South Africa in 1950, where he was educated at Athlone and Alexandra High Schools. In 1968, he obtained a university entrance with high marks in Physical Science and Mathematics. After working for some years in the computer industry, he completed his B.Sc. degree in Mathematics at the University of Natal. Later, he obtained a variety of teaching and computer-related qualifications, culminating in his Master of Commerce Degree (Information Systems), awarded with distinction by the University of Cape Town, in 1989. Since then, he has taught Information Systems at the Universities of Cape Town and Natal, and is currently a senior lecturer in Information Systems at the Northland Polytechnic in New Zealand. His particular research interest is the use of psychometric instruments to determine aspects of information system success. He has presented papers on this and related topics at conferences in South Africa, New Zealand and the United States.

**Marlon Parker** has spent the last three years working in the field of Information Systems and Business Intelligence. He served some time as a user and as an IT professional. Recently he has specialized in the area of technology-enhanced e-learning and assisted with the implementation of some online courses at the Cape

Copyright © 2003, Idea Group Inc. Copying or distributing in print or electronic forms without written permission of Idea Group Inc. is prohibited.

Technikon. He holds a B. Tech. Degree in Information Technology at the Cape Technikon.

**Shaun Pather** has over ten years of experience in IT training and education. Currently he is a Senior Lecturer in Information Systems, at the Cape Technikon. His teaching interests include Systems Analysis and Design, and Strategic Management Information Systems. He has been involved in various graduate and post-graduate IS programs for a variety of institutes, as well as executive management training. He has experience with consulting with various SMEs and IT training companies. His research interest is in the field of e-commerce evaluation.

**Paul A. Pavlou** is a Ph.D. Candidate of Information Systems at the Marshall School of Business in the University of Southern California (USC). He holds a master's degree in Electrical Engineering from USC, and a bachelor's degree in Managerial Studies and Electrical Engineering from Rice University (magna cum laude). His current research interests are mostly in the area of electronic commerce, dealing with interorganizational and consumer relationships, the role of trust and marketing communications. Mr. Pavlou is the author of several papers that appeared as journal articles, book chapters and conference proceedings.

**Sushil K. Sharma** is currently Assistant Professor of Information Systems at the Ball State University, Muncie, Indiana. He received his Ph.D. in Information Systems from Pune University, India and taught at the Indian Institute of Management, Lucknow for eleven years before joining Ball State University. Prior to joining Ball State, Dr. Sharma held the position of Visiting Research Associate Professor at the Department of Management Science, University of Waterloo, Canada. Dr. Sharma's primary teaching interests are e-commerce, computer communication networks, database management systems, management information systems, and information systems analysis and design. He has extensive experience in providing consulting services to several government and private organizations, including World Bank funded projects in the areas of information systems, e-commerce, and knowledge management. Dr. Sharma is the author of two books and has numerous articles in national and international journals. His current research interests include database management systems, networking environments, electronic commerce (e-commerce), knowledge management, and corporate information systems.

**Shawren Singh** is a lecturer at the University of South Africa (Unisa) in the Department of Computer Science and Information Systems. He is actively involved

Copyright © 2003, Idea Group Inc. Copying or distributing in print or electronic forms without written permission of Idea Group Inc. is prohibited.

in research relating to e-commerce, human computer interaction (HCI), Internet security, Web-based courseware tools, Internet applications, Web-based education and accounting information systems. He is a member of SAICSIT (South African Institute for Computer Scientists and Information Technologists, SAICSIT council member), SACLA (South African Computer Lecturers Association) and a member of the Unisa HCI Focus group. He has been invited as a guest speaker to local South African academic institutions and has presented his research at international conferences. He has lectured the following undergraduate level courses, ISD (Information Systems Development), Object-Oriented Approaches, Introduction to Computer Systems and Human-Computer Interaction.

**Charles A. Snyder** is the Woodruff Endowed Professor of MIS in the Department of Management at the College of Business, Auburn University. He received a Ph.D. in management from the University of Nebraska. He holds an M.S. in Economics from South Dakota State University, an M.B.A. from Ohio State University, and a B.F.A. from the University of Georgia. His research has appeared in the *Journal of Management Information Systems, Information & Management, The Academy of Management Review*, and many other leading journals. His research interests include knowledge management, information resource management, expert systems, computer-integrated manufacturing, systems analysis and design, and telecommunications management.

**Paul A. Taylor** is currently at the University of Leeds, UK. Previously, he was a senior lecturer in the sociology of technology at the University of Salford, Greater Manchester, UK. His main research interest centers upon the computer underground. In addition to numerous international articles, his main previous publication in this area is *Hackers: Crime in the Digital Sublime* (Routledge, 1999). He is currently working on several other book projects including another forthcoming Routledge book co-authored with Tim Jordan of the Open University and which is provisionally entitled "Hacktivists: Rebels with a Cause?"

**Roberto Vinaja** is an Assistant Professor of Computer Information Systems in the Department of Computer Information Systems and Quantitative Methods, University of Texas Pan American. His research interests include Global Electronic Commerce, Networking and Telecommunications.

Copyright © 2003, Idea Group Inc. Copying or distributing in print or electronic forms without written permission of Idea Group Inc. is prohibited.

# Index

## A

absenteeism 191
accreditation 247
active server pages 221
adaption-innovation theory 187, 204
adaptive problem-solvers 185
adverse selection 241
analyst-user dyad 203
analyst-user interface 204
anticipated continuity 250
artificial intelligence 99
asset structure 85
atmosphere 64

## B

B2B exchange 240, 246
benevolence 244
best-fitting regression 203
binding agent 127
biopolitical power 19
bit tax 39
brick and mortar 96
build-up 64
business risk 85
business-to-business (B2B) 51, 104
business-to-business (B2B) e-commerce 215
business-to-consumer (B2C) 51, 104, 215
business-to-employee (B2E) 51, 104

## C

capitalization 146
channels of communication 164
circulation of struggles 7
citizens-as-consumers 14
clustering strategy 5
cognitive processes 245
commercial tie-up 6
communication 64
computer anxiety 220
Computer Matching Privacy and Protection Act 37
computer-based information systems (CBIS) 52
computer-student 156
cookies 45
counter-populating 10
covert resistance 190
crack down 42
credibility 244
curates 13
customer-orientated 7
cyber slacking 42

## D

data mining 100
data privacy 215
data protection directive 216
data warehouses 100
Daz Doorstep Challenge 6
developing countries (DC) 23

Copyright © 2003, Idea Group Inc. Copying or distributing in print or electronic forms without written permission of Idea Group Inc. is prohibited.

devil's advocate stance  2
digital age  45
digital divide  36, 65
digital zapatismo  16
direct mail marketing  42
discipline  64
dot-com dreamers  140

**E**

e-commerce  23, 136, 213, 220
e-commerce accepted practice (EDAP)  50
e-commerce enterprises  137
e-commerce fiascoes  141
e-learning  160
economic development  25
economic transaction histories  89
educational technology (ET)  157
electronic civil disobedience  11
electronic commerce for developing
        countries  24
electronic data interchange (EDI)  111
electronic disturbance theatre (EDT)  11
electronic fund transfers (EFT)  54, 111
enterprise information systems  107
espace quelconque  5
EU (European Union)  24
everyone's information system  107
executive information systems (EIS)  103,
        107
executive support systems (ESS)  107

**F**

familiarity trust  243
feedback  147, 247
financial leverage  85
financial structure  85
flexibility  161
Free Trade Agreement between Canada,
        U.S.A, and Mexico  14
frequently asked questions (FAQ)  97
friction-free capitalism  2

**G**

generalised audit software (GAS) technique
        59

global information infrastructure (GII)  24
goodness-of-fit  196
government regulation  214
groupware  111
gung-ho frontier  3

**H**

hacktivists  18
high-penalty overhead  201
higher-risk tool  205
human dimension  163
hypertext markup language  221

**I**

impersonal trust  240, 243
incentives  241
information asymmetry  240
information systems (IS)  103
information technology (IT)  52
information warfare  36
infrastructure development  25
innovative analyst  194
innovators  185
inter-communication  162
International Telecommunication Union
        (ITU)  24
Internet relay chat (IRC)  160
Internet service providers (ISP)  74
interorganizational information system
        (IOIS)  246
ITQuadrant  141

**J**

Johannesburg Stock Exchange (JSE)  60
Just in Time (JIT)  127

**K**

Kirton Adaption-innovation Inventory
        (KAI)  185

**L**

laissez-faire ideology  4
legal bonds  249
leverage  85

Copyright © 2003, Idea Group Inc. Copying or distributing in print or electronic forms without written permission of Idea Group Inc. is prohibited.

low mean values  171

# M

management  147
manifest content  145
manifest destiny  3
marketing  147
McDonaldization  4
modus operandi  13
monitoring  248
monopolistic trends  40
monopoly practices  39
moral hazard  241
multi-national cooperation  25
multiple constituency approach  109

# N

National Information Infrastructure (NII)  24
neo-tribes  8
networked computing  112
new economy  34
new labour  6

# O

online analytical processing (OLAP)  106
online customer service  96
operating leverage  85
operation turnout  6
operational objectives  73
opportunism  241
organizational decision makers  89
organizational information visualization  88
Organization for Economic Cooperation
    and Development (OECD)  24
overt resistance  190

# P

paradox of incommunicability  7
parameters  148
peer-to-peer  167
perceived risk  249
personal information  215
personal information privacy  215
political democratization  24
precise contradiction  204

pricing  250
privacy  213
privacy invasion  219,  222
privacy policies  213
propensity to purchase  222
psychological withdrawal  191
pure intellectual distance  18

# R

r-score  196
race towards weightlessness  6
regulatory preference  213
return on investment (ROI)  70

# S

satisfaction  249
security  148
self-regulation  229
semiological guerilla warfare  11
shared data bases  112
shared mindset  72
shoulder-looking  42
signals  241
small, medium and micro enterprises
    (SMMEs)  63
small to medium enterprises (SMEs)  122
social drama  16
social isolation  43
social obsolescence  77
South African National Small Business Act
    122
South African revenue services (SARS)  57
South African SME business sector  128
supply chain  213
supply chain management (SCM)  112, 143
surrogate  190
system satisfaction schedule  195
systems development life cycle (SDLC)  139

# T

teacher-computer-student  156
teacher-student  156
technology dimension  163
technology knowledge  219
technology objectives  73

Copyright © 2003, Idea Group Inc. Copying or distributing in print or electronic forms without written permission of Idea Group Inc. is prohibited.

technology-delivered e-learning environ-
    ment 173
tertiary institutions 159
text conferencing 98
three-ring binder 4
trust 243
trust-building cognitive processes 241

**U**

uncertainty 240
underground puissance 8
universal tool 158

**V**

value chain integration 127
value-added networks (VANs) 111
vertical events 7
videoconferencing session 99
virtual classroom 159
virtual learning environments (VLEs) 159
voice over the Internet protocol (VOIP) 97

**W**

web surfing 35
web-based Instructional Systems model
    164
wireless application protocol (WAP) 105
World Wide Web (WWW) 53, 172

Copyright © 2003, Idea Group Inc. Copying or distributing in print or electronic forms without written
permission of Idea Group Inc. is prohibited.

# InfoSci-Online
# Database

**30-Day free trial!**

## www.infosci-online.com

Provide instant access to the latest offerings of Idea Group Inc. publications in the fields of INFORMATION SCIENCE, TECHNOLOGY and MANAGEMENT

During the past decade, with the advent of telecommunications and the availability of distance learning opportunities, more college and university libraries can now provide access to comprehensive collections of research literature through access to online databases.

The InfoSci-Online database is the most comprehensive collection of *full-text* literature regarding research, trends, technologies, and challenges in the fields of information science, technology and management. This online database consists of over 3000 book chapters, 200+ journal articles, 200+ case studies and over 1,000+ conference proceedings papers from IGI's three imprints (Idea Group Publishing, Information Science Publishing and IRM Press) that can be accessed by users of this database through identifying areas of research interest and keywords.

**Contents & Latest Additions:**
Unlike the delay that readers face when waiting for the release of print publications, users will find this online database updated as soon as the material becomes available for distribution, providing instant access to the latest literature and research findings published by Idea Group Inc. in the field of information science and technology, in which emerging technologies and innovations are constantly taking place, and where time is of the essence.

The content within this database will be updated by IGI with 1300 new book chapters, 250+ journal articles and case studies and 250+ conference proceedings papers per year, all related to aspects of information, science, technology and management, published by Idea Group Inc. The updates will occur as soon as the material becomes available, even before the publications are sent to print.

InfoSci-Online pricing flexibility allows this database to be an excellent addition to your library, regardless of the size of your institution.

**Contact: Ms. Carrie Skovrinskie, InfoSci-Online Project Coordinator, 717-533-8845 (Ext. 14), cskovrinskie@idea-group.com for a 30-day trial subscription to InfoSci-Online.**

**A product of:**

## INFORMATION SCIENCE PUBLISHING*
Enhancing Knowledge Through Information Science
http://www.info-sci-pub.com

*an imprint of Idea Group Inc.*

# Journal of Electronic Commerce in Organizations (JECO)

## The International Journal of Electronic Commerce in Modern Organizations

Journal of Electronic Commerce in Organizations

**ISSN:** 1539-2937
**eISSN:** 1539-2929
**Subscription:** Annual fee per volume (4 issues):
Individual US $85
Institutional US $185

**Editor:** Mehdi Khosrow-Pour, D.B.A.
Information Resources
Management Association, USA

## Mission

The *Journal of Electronic Commerce in Organizations* is designed to provide comprehensive coverage and understanding of the social, cultural, organizational, and cognitive impacts of e-commerce technologies and advances on organizations around the world. These impacts can be viewed from the impacts of electronic commerce on consumer behavior, as well as the impact of e-commerce on organizational behavior, development, and management in organizations. The secondary objective of this publication is to expand the overall body of knowledge regarding the human aspects of electronic commerce technologies and utilization in modern organizations, assisting researchers and practitioners to devise more effective systems for managing the human side of e-commerce.

## Coverage

This publication includes topics related to electronic commerce as it relates to: Strategic Management, Management and Leadership, Organizational Behavior, Organizational Developement, Organizational Learning, Technologies and the Workplace, Employee Ethical Issues, Stress and Strain Impacts, Human Resources Management, Cultural Issues, Customer Behavior, Customer Relationships, National Work Force, Political Issues, and all other related issues that impact the overall utilization and management of electronic commerce technologies in modern organizations.

**For subscription information, contact:**

**Idea Group Publishing**
701 E Chocolate Ave., Ste 200
Hershey PA 17033-1240, USA
cust@idea-group.com
URL: www.idea-group.com

**For paper submission information:**

**Dr. Mehdi Khosrow-Pour**
**Information Resources Management**
**Association**
jeco@idea-group.com